Tauberian Theory of Wave Fronts in Linear Hereditary Elasticity

Alexander A. Lokshin

Tauberian Theory of Wave Fronts in Linear Hereditary Elasticity

 Springer

Alexander A. Lokshin
Department of Mathematics and Informatics in Primary School
Moscow Pedagogical University
Moscow, Russia

ISBN 978-981-15-8580-7   ISBN 978-981-15-8578-4   (eBook)
https://doi.org/10.1007/978-981-15-8578-4

This Springer imprint is published by the registered company Springer Nature Singapore Pte Ltd.
The registered company address is: 152 Beach Road, #21-01/04 Gateway East, Singapore 189721, Singapore

Preface

This book is devoted to convolution type equations, which occur in linear wave problems of hereditary elasticity. The main mathematical tool used below is the Fourier–Laplace transform.

The possibility of making use of the Fourier–Laplace transform, when solving convolution type equations, is evident. After applying the mentioned transform, we get a simple formula for the transformed solution of the equation considered. However the essence of matter is to derive from this formula the description of behavior of the solution itself: to find its support, asymptotics, etc. We must note that, as a rule, the Fourier–Laplace transform is used only formally in papers on wave problems of hereditary elasticity, whereas profound mathematical theorems (such as the Paley–Wiener theorem and Tauberian theorems) are neglected.

The purpose of this book is to construct a rigorous mathematical approach to linear hereditary problems of wave propagation theory and to demonstrate usefulness of profound mathematical theorems in hereditary mechanics.

Chapter 1 is introductory, which contains some preliminary material from linear hereditary elasticity, geometry, and harmonic analysis. Chapter 2 investigates conditions of hyperbolicity for general operators with memory in case of many spatial variables. Operators of such a kind occur in problems of wave propagation in anisotropic hereditary media. It turns out that under certain geometrical restrictions of monotonicity and concavity type imposed on the functions of memory, the desired hyperbolicity condition can be formulated algebraically. Chapter 3 deals with more refined properties of wave equations with memory. This chapter discusses the one-dimensional case, which corresponds to wave propagation in linearly elastic hereditary rods. By using both real end complex Tauberian techniques, a classification of near-front asymptotics of solutions to equations considered can be given, depending on the singularity character of the memory function. In particular, it is rigorously demonstrated that in linear hereditary media with singular memory (i.e., with a memory function tending to infinity when approaching the current moment of time), strong shocks cannot propagate at all. Among other results of Chap. 3, a mathematically rigorous derivation of the formula is mentioned for wave front velocity in an inhomogeneous hereditary rod with singular memory (Sect. 3.9) and a generalization of the well-known Cagniard–de Hoop method to the hereditary case (Sect. 3.10). The last two results demonstrate the importance of nonlinear Laplace transform in linear hereditary elasticity.

Department of Mathematics and   Alexander A. Lokshin
Informatics in Primary School
Moscow Pedagogical University
Moscow, Russia
April 1994–June 2020

Notation

1. $\mathbb{R}^n$: n-dimensional real Euclidean space
2. $\mathbb{C}^n$: n-dimensional complex Euclidean space
3. The Laplace transform of a function $f(y)$, $y \in \mathbb{R}^1$,

$$L_{y \to p} \equiv \int_0^\infty f(y)e^{-py}\mathrm{d}y, p \in \mathbb{R}^1.$$

Here the integral is supposed to be Lebesgue convergent for $p > 0$ large enough.

4. The Fourier-Laplace transform of a function $f(y)$, $y \in \mathbb{R}^n$,

$$F_{y \to z} \equiv \int_{-\infty}^\infty \cdots \int_{-\infty}^\infty f(y)e^{-izy}\mathrm{d}y_1 \cdots \mathrm{d}y_n,$$

where

$$z \in \mathbb{C}^n, \quad zy \equiv z_1 y_1 + \ldots + z_n y_n$$

Here the integral is supposed to be Lebesgue convergent when Im z is contained in some open set in Im $\mathbb{C}^n$.

5. For the Laplace transform and the Fourier–Laplace transform with respect to $t \in \mathbb{R}^1$ we use special notation:

$$L_{t \to p}f \equiv \bar{f}(p), \quad p \in \mathbb{R}^1$$

$$F_{t\to\lambda}f \equiv \widetilde{f}(\lambda), \quad \lambda \in \mathbb{C}^1$$

6. The convolution with respect to t:

$$f(t) * g(t) \equiv \int\limits_{-\infty}^{\infty} f(t-\tau)\,g(\tau)\mathrm{d}\tau$$

7. In Sects. 2.2 and 2.3, $\overline{\lambda}$, $\overline{E}$, $\overline{Z}$ denote the complex conjugate of λ, E, Z

Contents

About the Author

Alexander A. Lokshin is Professor of Mathematics in the Faculty of Primary Education, Moscow Pedagogical University, Russia, since 1999. He completed his graduation in differential equations in 1973 in the Faculty of Mechanics and Mathematics, Moscow State University, Russia. Professor Lokshin defended his thesis on "On lacunas and weak lacunas of hyperbolic and quasi-hyperbolic equations" in 1976 at Moscow State University, Russia. Later, he defended his doctoral dissertation on "Waves in hereditarily elastic media" at the Institute of Problems of Mechanics, USSR Academy of Sciences, Russia, in 1985. He served as a junior research fellow at the Moscow Institute of Electronic Engineering and had also worked as a scientific editor for the Moscow University Press, Russia. Coauthor of *The Mathematical Theory of Wave Propagation in Media with Memory* and *Nonlinear Waves in Inhomogeneous and Hereditary Media,* Prof. Lokshin has also published several books proposing a visual and at the same time mathematically rigorous justification of the four arithmetic algorithms.

Chapter 1
The Problem of Hyperbolicity in Linear Hereditary Elasticity

This chapter discusses the problem of hyperbolicity in linear hereditary elasticity. In Sect. 1.1, preliminary information about integro-differential dynamic operators of hereditary elasticity is presented. Sections 1.2, 1.3, and 1.4 are devoted to some well-known facts from geometry and harmonic analysis. Section 1.5 contains lemmas concerning the Fourier–Laplace transform of the function of memory.

1.1 Dynamic Problems of Hereditary Elasticity and Integro-Differential Operators

1.1.1 One-Dimensional Constitutive Equations

Let a homogeneous linear hereditary rod be located on the x-axis. We suppose stress σ and deformation ε of the material of the rod to be related by the following constitutive equation

$$\sigma(t,x) = A\left[\varepsilon(t,x) - \int_{-\infty}^{t} R(t-\tau)\varepsilon(\tau,x)\mathrm{d}\tau\right]$$

Here t is time, A (a constant) is the instantaneous module of elasticity, and the function $R(t)$ defined for $t > 0$ is the relaxation kernel. Let us extend $R(t)$ to the semi-axis $t < 0$ by zero. Then the previous relation can be rewritten as

$$\sigma = A[\varepsilon - R(t) * \varepsilon] \tag{1.1.1}$$

A. A. Lokshin, *Tauberian Theory of Wave Fronts in Linear Hereditary Elasticity*, https://doi.org/10.1007/978-981-15-8578-4_1

where * denotes convolution with respect to t over the whole of t-axis. Let us single out in (1.1.1) the operator applied to ε. Then Eq. (1.1.1) takes the form

$$\sigma = A[1 - R(t)*]\varepsilon \qquad (1.1.1')$$

Now, we solve (1.1.1') for ε by multiplying both sides of this equation by the operator $1/[1 - R(t)*]$, which is a Taylor's series $1 + R(t)* + R(t) * R(t) * + \cdots$. Now, we have

$$\varepsilon = \frac{1}{A} \frac{1}{1 - R(t)*} \sigma$$

Now, we denote

$$\frac{1}{1 - R(t)*} \equiv 1 + \Lambda(t)* \qquad (1.1.2)$$

The operator equality (1.1.2) yields two equivalent relations which link the functions $R(t)$ and $\Lambda(t)$. Namely

$$\Lambda(t) = R(t) + R(t) * R(t) + \cdots \qquad (1.1.3)$$

whence, **in** particular, it follows that $\Lambda(t) = 0$ for $t < 0$, and the relation dual to (1.1.3)

$$R(t) = \Lambda(t) - \Lambda(t) * \Lambda(t) + \cdots \qquad (1.1.4)$$

The function $\Lambda(t)$, defined by (1.1.2), is called the *creep kernel*. Using $\Lambda(t)$, *we can finally* rewrite the expression for deformation by means of stress in the form

$$\varepsilon = \frac{1}{A} [\sigma + \Lambda(t) * \sigma] \qquad (1.1.5)$$

1.1.2 Regular and Singular Functions of Memory

One can easily see that, under reasonable restrictions, (1.1.3), or (1.1.4), yields

$$R(t) \sim \Lambda(t) \text{ as } t \to +0$$

Functions of memory tending to a finite limit as $t \to +0$ are called *regular*, while functions of memory tending to infinity as $t \to +0$ are called *singular*. Thus the kernels $R(t)$ and $\Lambda(t)$ are either regular or singular simultaneously.

One may be taking singularity of memory functions as an exception to the rule. However, it is not the case in continuum mechanics. On the contrary, singular kernels appear rather often. The presence of singularity is corroborated by specific high-frequency experiments and also by theoretical considerations (see, e.g. [2]).

Mathematically, the regular case (easier for investigation) is studied in detail [3, 4]. So, this case is of no interest for us. In this book, we only consider singular functions of memory. But, it should be noted that all the results concerning regular functions of memory can be easily derived from our theory.

1.1.3 One-Dimensional Wave Operators with Memory

Here, we shall try to construct a general theory of linear operators describing wave propagation in homogeneous hereditary elastic media. One of the most important operators of this theory, the wave operator with memory, appears in the following situation.

Let us turn to the rod which is under consideration in Sect. (1.1.1). For simplicity, we suppose the rod to be infinite and its density is ρ (a constant). Let the rod be subjected to the longitudinal external load $f(t, x)$ counted at a unit length. *Then* the equation of the rod motion will obviously take the form

$$\rho \frac{\partial^2 u}{\partial t^2} = \frac{\partial \sigma}{\partial x} + f(t, x). \tag{1.1.6}$$

Here u is the displacement of the element of the rod. Please note that the displacement u and deformation ε are linked by formula

$$\varepsilon = \frac{\partial u}{\partial x} \tag{1.1.7}$$

Substituting the expression for σ given by (1.1.1) into (1.1.6) and taking into account (1.1.7), we arrive at the one-dimensional wave equation with memory as below:

$$\frac{\partial^2 u}{\partial t^2} - c^2 \left[\frac{\partial^2 u}{\partial x^2} - R(t) * \frac{\partial^2 u}{\partial x^2} \right] = \frac{f}{\rho}, \quad c \equiv \sqrt{\frac{A}{\rho}} \tag{1.1.8}$$

Applying the operator $1/[1 - R(t)*]$ to both sides of (1.1.8) and taking into account (1.1.2), we arrive at the following equivalent form:

$$[1 + \Lambda(t)*]\frac{\partial^2 u}{\partial t^2} - c^2 \frac{\partial^2 u}{\partial x^2} = [1 + \Lambda(t)*]\frac{f}{\rho} \qquad (1.1.9)$$

Operators from the left-hand sides of (1.1.8) and (1.1.9) are called *one-dimensional wave operators with memory.*

1.1.4   Wave Operators with Memory in Cases of Two and Three Spatial Dimensions

Let us consider equations of motion of an unbounded (two or three dimensional) homogeneous linear hereditary medium of density ρ:

$$\rho \frac{\partial^2 \vec{u}}{\partial t^2} = \{\lambda[1 - q(t)*] + 2\mu[1 - h(t)]\} \nabla \left(\nabla \cdot \vec{u} \right)$$
$$-\mu[1 - h(t)*]\nabla \times \left(\nabla \times \vec{u} \right) + \vec{f}(t,x) \qquad (1.1.10)$$

Here $\vec{u} = (u1, \ldots, un)$ is the vector field of displacement; $\nabla = \left(\frac{\partial}{\partial x_1}, \ldots, \frac{\partial}{\partial x_n} \right)$; $\lambda > 0$ and $\mu > 0$ are the instantaneous elastic constants of Lamé; $q(t)$ and $h(t)$ are the corresponding relaxation kernels; $\vec{f} = (f_1, \ldots, f_n)$ is the body force; and $n = 2$ or 3.

Let us represent the force $\vec{f}$ by means of the potentials of Helmholtz:

$$\vec{f} = \nabla \Phi + \nabla \times \vec{\Psi}, \quad \nabla \cdot \vec{\Psi} = 0$$

Then, by the theorem of Lamé (see 4.5), for $\vec{u}$ there exist potentials ϕ and ψ, such that

1. $\vec{u} = \nabla \phi + \nabla \times \vec{\psi}$;

2. $\nabla \cdot \vec{\psi} = 0$;

3. $\dfrac{\partial^2 \phi}{\partial t^2} - \dfrac{\lambda[1 - q(t)*] + 2\mu[1 - h(t)*]}{\rho} \Delta\phi = \dfrac{\Phi}{\rho}$;   (1.1.11)

4. $\dfrac{\partial^2 \vec{\psi}}{\partial t^2} - \dfrac{\mu[1 - h(t)*]}{\rho} \Delta\vec{\psi} = \dfrac{\vec{\Psi}}{\rho}$ (1.1.12)

Thus, the left-hand sides of (1.1.11) and (1.1.12) contain wave operators with memory in case of two or three spatial dimensions. In Chap. 3, we shall get

acquainted with one more situation where the two-dimensional wave operator with
memory appears.

1.1.5 Anisotropic Hereditary Elastic Medium: Systems of Dynamic Equations and Their Determinants

Consider an unbounded homogeneous anisotropic hereditary elastic medium. Let the
dimension of the medium be equal to 2 or 3, as is well-known, for such a medium the
relation between stress tensor σ_{ik} and deformation tensor ε_{lm} is given by the formula

$$\sigma_{ik} = [\lambda_{iklm} + g_{iklm}(t)*]\varepsilon_{lm} \tag{1.1.13}$$

Here λ_{iklm} are the instantaneous elastic moduli, $g_{iklm}(t)$ the corresponding memory functions which are supposed to be equal zero for $t < 0$. In addition, the
equalities

$$\left.\begin{array}{l} \lambda_{iklm} = \lambda_{kilm} = \lambda_{ikml} = \lambda_{lmik} \\ g_{iklm}(t) = g_{kilm}(t) = g_{ikml}(t) = g_{lmik}(t) \end{array}\right\} \tag{1.1.14}$$

hold. We would like to recall here that deformation tensor ε_{lm} and displacement $\vec{u}$
are related by the equality

$$\varepsilon_{lm} = \frac{1}{2}\left(\frac{\partial u_l}{\partial x_m} + \frac{\partial u_m}{\partial x_l}\right) \tag{1.1.15}$$

Now, let us write equations of motions of the medium:

$$\rho\frac{\partial^2 u_i}{\partial t^2} = \frac{\partial^2 \sigma_{ik}}{\partial x_k} + f_i(t,x); \quad i = 1, \ldots, n \tag{1.1.16}$$

Here $\vec{f} = (f_1, \ldots, f_n)$ is the body force. For simplicity, the functions $f_i(t, x)$ are
supposed to have a compact support. Inserting expression (1.1.13) for σ_{ik} into the
last system of equations and taking into account (1.1.14) and (1.1.15), we obtain the
equations of motions of the medium in terms of displacement:

$$\frac{\partial^2 u_i}{\partial t^2} - \frac{1}{\rho}[\lambda_{iklm} + g_{iklm}(t)*]\frac{\partial^2 u_m}{\partial x_k \partial x_l} = \frac{f_i(t,x)}{\rho} \tag{1.1.17}$$

Now suppose all the memory functions $g_{iklm}(t)$ can be represented in the form

$$g_{iklm}(t) = c_{iklm}\phi(t) \tag{1.1.18}$$

where c_{iklm} are some constants and the function $\phi(t)$ is equal to zero for $t < 0$. Then elements of a matrix $G = \|G_{ij}\|$ of the system (1.1.17) can be represented as follows:

$$G_{ij} = G_{ij0}\left(\frac{\partial}{\partial t}, \frac{\partial}{\partial x}\right) + \phi(t) * G_{ij1}\left(\frac{\partial}{\partial t}, \frac{\partial}{\partial x}\right)$$

where G_{ij0}, G_{ij1} are linear homogeneous second-order differential operators with constant coefficients. Since differential operators with constant coefficients can be considered operators of convolution (with δ function and its derivatives), elements of the matrix G can be also considered operators of convolution (with respect to t and x).

Now, let us successively eliminate all the unknown functions, except some u_k, from the system (1.1.17). Evidently, all our operations will be similar to the ones in linear algebra. The only difference lies in the fact that we apply convolution operators to the equations of the system under consideration, instead of multiplying them by numbers. In doing this, we essentially make use of commutativity of convolution operators.

As a result, we arrive at the following equalities corresponding to the Cramer's formulas from linear algebra:

$$(\det G)u_k = \det B^k; \quad k = 1, \ldots, n. \tag{1.1.19}$$

Here $B^k = \left\|B_{ij}^k\right\|$ is a matrix, which results from the matrix G after substituting its k-th column for the column

$$\begin{pmatrix} f_1/\rho \\ \vdots \\ f_n/\rho \end{pmatrix};$$

$\det G$ and $\det B^k$ denote the corresponding matrix determinants taken with respect to operation of convolution; all the f_i/ρ should thus be when calculating $\det B^k$ treated as convolution operators with the kernels $f_i(t, x)/\rho$. Finally, the expression $(\det G) u_k$ denotes the operator $\det G$ which is applied to the function u_k. It is easy to see that in the three-dimensional case

$$\det G = V_0\left(\frac{\partial}{\partial t},\frac{\partial}{\partial x}\right) + \phi(t) * V_1\left(\frac{\partial}{\partial t},\frac{\partial}{\partial x}\right) + \phi(t) * \phi(t) * V_2\left(\frac{\partial}{\partial t},\frac{\partial}{\partial x}\right)$$

$$+ \phi(t) * \phi(t) * \phi(t) * V_3\left(\frac{\partial}{\partial t},\frac{\partial}{\partial x}\right) \tag{1.1.20}$$

where V_s are homogeneous sixth order linear differential operators with constant coefficients. In the two-dimensional form, the convolutional determinant of the matrix G has the following form:

$$\det G = V_0\left(\frac{\partial}{\partial t},\frac{\partial}{\partial x}\right) + \phi(t) * V_1\left(\frac{\partial}{\partial t},\frac{\partial}{\partial x}\right) + \phi(t) * \phi(t)$$

$$* V_2\left(\frac{\partial}{\partial t},\frac{\partial}{\partial x}\right) \tag{1.1.21}$$

where V_s are homogenous fourth-order linear differential operators with constant coefficients.

Note It can be demonstrated that in the isotropic case det G can be decomposed into a product of wave operators with memory. Namely, for $n = 3$

$$\det G = \left[\frac{\partial^2}{\partial t^2} - \frac{\lambda(1 - q(t)*) + 2\mu(1 - h(t)*)}{\rho}\Delta\right]$$

$$\times \left[\frac{\partial^2}{\partial t^2} - \frac{\mu(1 - h(t)*)}{\rho}\Delta\right]^2 \tag{1.1.22}$$

where $\Delta \equiv \Sigma_{i=1}^{3}\frac{\partial^2}{\partial x_i^2}$, and for $n = 2$

$$\det G = \left\{\frac{\partial^2}{\partial t^2} - \frac{\lambda[1 - q(t)*] + 2\mu[1 - h(t)*]}{\rho}\Delta\right\}$$

$$\times \left\{\frac{\partial^2}{\partial t^2} - \frac{\mu[1 - h(t)*]}{\rho}\Delta\right\}, \tag{1.1.23}$$

where $\Delta \equiv \Sigma_{i=1}^{2}\frac{\partial^2}{\partial x_i^2}$. Here, on account of the hypothesis 1.1.17

$$q(t) = \text{const}_1\phi(t), \quad h(t) = \text{const}_2\phi(t)$$

1.1.6   The Problem of Hyperbolicity

The question of interest (both from the mathematical and physical points of view) is whether dynamic equations of linear hereditary elasticity have solutions belonging to reasonable classes of functions and describing finite speed wave propagation. For example, let us consider Eq. (1.1.8):

$$\frac{\partial^2 u}{\partial t^2} - c^2 \left[\frac{\partial^2 u}{\partial x^2} - R(t) * \frac{\partial^2 u}{\partial x^2} \right] = \frac{f(t, x)}{\rho} \qquad (1.1.24)$$

For simplicity, we suppose $f(t, x)$ to have a compact support in the t, x-plane. Furthermore, let $E(t, x)$ be some fundamental solution for the operator (1.1.9). That is, a solution of the equation

$$\frac{\partial^2 E(t, x)}{\partial t^2} - c^2 \left[\frac{\partial^2 E(t, x)}{\partial x^2} - R(t) * \frac{\partial^2 E(t, x)}{\partial x^2} \right] = \delta(t)\delta(x) \qquad (1.1.25)$$

Then the convolution

$$u(t, x) = \int\limits_{-\infty}^{\infty} \int\limits_{-\infty}^{\infty} E(t - \tau, x - \xi)\frac{f(\tau, \xi)}{\rho}\, d\tau d\xi$$

will evidently satisfy Eq. (1.1.24). One can easily see that for the convolution (1.1.25) to describe the finite speed wave propagation from the source $f(t, x)$, the fundamental solution $E(t, x)$ must satisfy the condition:

$$\text{supp}E(t, x) \subseteq \{t, x | t \geq \text{const}|x|\}, \text{const} > 0 \qquad (1.1.27)$$

Here supp $E(t, x)$ denotes the support of the fundamental solution $E(t, x)$. But does there exist such a fundamental solution for the operator (1.1.9)? Let us suppose now $x \in \mathbb{R}^n$, $n > 1$, and let $|x| = (x^2_1 + \cdots + x_n^2)^{1/2}$.

Definition Suppose an operator P has a fundamental solution $E(t, x)$ satisfying the condition (1.1.27). Then we shall call the operator P hyperbolic in D' (or, if it will not lead to misunderstanding, simply hyperbolic). Here D' is the space of distributions which are continuous linear functionals on the space D of infinitely differentiable test functions with compact support [6].

Definition Suppose an operator P has a fundamental solution $E(t, x)$ such that $e^{-Mt}E(t, x) \in S'$ for some $M > 0$ and the condition (1.1.27) holds. Then we shall call the operator P hyperbolic in S'. Here S' is the space of temperate distributions, which are continuous linear functionals on the space S of rapidly decreasing test functions [6]. It is clear that hyperbolicity in D' follows from hyperbolicity in S'

(since $S' \subset D'$). The space S' is better adapted for deriving necessary and sufficient conditions of hyperbolicity. Therefore, our main results will concern operators hyperbolic in S'. Some results will be also established in D'.

Note As is well-known, the differential wave operator is hyperbolic (in D'). The sixth-order differential operator V_0 in (1.1.20) and the fourth-order differential operator V_0 in (1.1.21) are also hyperbolic. These operators correspond to the instantaneous elastic behaviour of the hereditary medium.

Example Let $k > 0$, $0 < \alpha < 1$. Then, as we shall see in Chap. 2, the wave operator with memory

$$\frac{\partial^2}{\partial t^2} - c^2 \left(\frac{\partial^2}{\partial x^2} - kt_+^{-\alpha} * \frac{\partial^2}{\partial x^2} \right);$$

where

$$t_+^{-\alpha} = \begin{cases} t^{-\alpha}, & \text{for } t > 0 \\ 0, & \text{for } t < 0 \end{cases}$$

is hyperbolic in S', while the operator

$$\frac{\partial^2}{\partial t^2} - c^2 \left(\frac{\partial^2}{\partial x^2} + kt_t^{-\alpha} \frac{\partial^2}{\partial x^2} \right)$$

is not hyperbolic in S' (and even in D').

This result, evidently, is in accordance with non-negativeness of the relaxation kernel. In fact, hyperbolicity in S' of the wave operator with singular memory is determined by the sign of the convolutional summand. Note, however, that for operators of a more general form, the algebraic criterion of hyperbolicity in S' is more interesting.

1.1.7 Equivalence of Hyperbolicity of the System of Dynamic Equations to Hyperbolicity of Its Convolutional Determinant

Let us return to the system (1.1.17) and (1.1.18)

$$\frac{\partial^2 u_i}{\partial t^2} - \frac{1}{\rho} [\lambda_{iklm} + c_{iklm}\phi(t)*] \frac{\partial^2 u_m}{\partial x_k \partial x_l} = \frac{f_i(t,x)}{\rho}, i = 1, \ldots, n; \qquad (1.1.28)$$

for conciseness, we rewrite it in the form

$$G\vec{u} = \frac{\vec{f}}{\rho} \tag{1.1.28'}$$

Here G denotes the matrix of the system (1.1.28):

$$\vec{u} = (u1, \ldots, un); \vec{f} = (f_1, \ldots, f_n); n = 2 \ or \ 3$$

We would remind to the reader that the matrix $E(t,x) = \|E_{ij}(t,x)\|$ is called the *fundamental matrix of solutions* for the system (1.1.28'), if $E(t, x)$ satisfies the matrix equation

$$GE(t,x) = \delta(t)\delta(x)I \tag{1.1.29}$$

where I is the unity matrix. We shall call the system (1.1.28') hyperbolic in D' (in S'), if there exists a fundamental matrix of solutions $E(t, x)$ for (1.1.28') such that

$$\operatorname{supp} E_{ij}(t,x) \subseteq \{t, x | t \geq \operatorname{const}|x|\}, \quad \operatorname{const} > 0, \tag{1.1.30}$$

and $E_{ij}(t,x) \in D'$ (respectively, $e^{-Mt}E(t,x) \in S'$ for some $M > 0$). Here

$$| x | = (x_1^2 + x_2^2 + \cdots + x_n^2)^{1/2}$$

The following proposition is similar to the one given in [7] for the purely differential case.

Proposition The system (1.1.28') is hyperbolic in D' (in S'), if and only if the operator $\det G$ is hyperbolic in D' (respectively, in S'). Here $\det G$ is the convolutional determinant of the matrix G.

Proof Let the system (1.1.28') be hyperbolic (e.g. in D'), and let $E(t,x) = \|E_{ij}(t,x)\|$ be the corresponding fundamental matrix of solutions describing finite speed wave propagation. Then one can take the convolutional determinant of the matrix equality (1.1.29), since all convolutions which enter $\det (GE)$ are evidently determined. Using the commutativity of convolution, we have

$$\det G \det E(t,x) = \delta(t)\delta(x) \tag{1.1.31}$$

The convolutional determinants $\det G$ and $\det E$ are, obviously, also determined. Furthermore, from (1.1.30), it follows that

$$\operatorname{supp} \det E(t,x) \subseteq \{t, x | t \geq \operatorname{const}|x|\}, \operatorname{const} > 0.$$

It is also clear that $\det E(t, x) \in D'$. Thus the operator $\det G$ proves to be hyperbolic in D'.

Conversely, let the operator det G be hyperbolic in D', and let E (det G) be its finite speed fundamental solution. Let $Q = \|Q_{ij}\|$ be the matrix associated with G, that is, $GQ = I \det G$. Then the desired finite speed fundamental matrix of solutions for (1.1.28') is given by formula

$$E(G) = QE(\det G)I$$

In fact, let us apply G to the last equality. We have

$$GE(G) = GQE(\det G) \cdot I = \det G \cdot I \cdot E(\det G) \cdot I = \delta(t)\delta(x)I$$

Therefore, $E(G)$ really is a fundamental matrix of solutions for Eq. (1.1.28'). Moreover, one can easily see that the elements Q_{ij} of the matrix Q have a structure similar to the one of det G. Therefore, the functions

$$E_{ij}(G) = Q_{ij}E(\det G)$$

satisfy Eq. (1.1.30). The fact that $E_{ij}(G) \in D'$ is obvious. The case of hyperbolicity in S' can be studied in the same manner.

Note It is easy to see that our approach can be generalized to the case where the kernels $g_{iklm}(t)$ in (1.1.13) can be represented as convergent series:

$$g_{iklm}(t) = c_{iklm1}\phi(t) + c_{iklm2}\phi(t) * \phi(t) + \cdots \tag{1.1.32}$$

In this case, det G also assumes the form of a series:

$$\det G = V_0\left(\frac{\partial}{\partial t}, \frac{\partial}{\partial x}\right) + \phi(t) * V_1\left(\frac{\partial}{\partial t}, \frac{\partial}{\partial x}\right) + \cdots \tag{1.1.33}$$

Let us explain how the operator (1.1.33) acts. Let $u(t, x)$, $f(t, x)$ be some distributions (e.g., from the space D'). Then the equality

$$\left[V_0\left(\frac{\partial}{\partial t}, \frac{\partial}{\partial x}\right) + \cdots + \phi(t) * \ldots * \phi(t) * V_n\left(\frac{\partial}{\partial t}, \frac{\partial}{\partial x}\right)\right]u(t, x) = f(t, x)$$

means that for each test function $\psi(t, x)$

$$\lim_{N \to \infty} < [V_0 + \cdots + \phi * \cdots \phi * V_N]u(t, x), \phi(t, x) > = < f(t, x), \phi(t, x) >$$

Note If in the isotropic case the memory functions satisfy (1.1.32), then it is evident that in (1.1.22) and (1.1.23), the kernels $q(t)$ and $h(t)$ also have the structure of the (1.1.32) type.

Note Suppose the operator det G can be represented as a product of operator multipliers. Furthermore, let these operator multipliers be hyperbolic in D' (or in S'). Then the operator det G will also be hyperbolic in D' (respectively, in S'). From the theorems given below, the inverse proposition also follows. Thus in the isotropic case of 2 or 3 spatial dimensions, hyperbolicity of wave operators with memory plays the role, which is as important as the one of hyperbolicity of the one-dimensional wave operator with memory.

1.1.8   Definition of the Class of Integro-Differential Operators Under Consideration

Here we shall study hyperbolicity of operators of the (1.1.33) type, which are the natural generalization of dynamic operators of hereditary elasticity. In what follows, we consider the number of spatial dimensions equal to n ($n \geq 1$).

Let us introduce the following notation for operators from the class under consideration:

$$W = V_0\left(\frac{\partial}{\partial t}, \frac{\partial}{\partial x}\right) + \phi(t) * V_1\left(\frac{\partial}{\partial t}, \frac{\partial}{\partial x}\right) + \phi(t) * \phi(t) * V_2\left(\frac{\partial}{\partial t}, \frac{\partial}{\partial x}\right)$$
$$+ \cdots;$$

(1.1.34)

and

$$V = V_0\left(\frac{\partial}{\partial t}, \frac{\partial}{\partial x}\right) + \phi(t) * V_1\left(\frac{\partial}{\partial t}, \frac{\partial}{\partial x}\right)$$

(1.1.35)

In Chaps. 1 and 2, we suppose that

$$\phi(t) = 0, \text{ for } t < 0;$$

1. $\phi(t)$ is four times differentiable for $t > 0$.
2. $(-1)^k \frac{d^k \phi(t)}{dt^k} \geq 0$, for $t > 0$, $k = 0, 1, 2$ and 3.
   $\phi(t) \rightarrow +\infty$, as $t \rightarrow +0$;
3. There exists $r \in (0, 1)$ such that function $t^r \phi(t)$ is increasing for small $t > 0$.

Besides that we suppose that

(a) $V_s = V_s\left(\frac{\partial}{\partial t}, \frac{\partial}{\partial x}\right); s = 0, 1, \ldots$ are homogeneous mth-order differential operators with constant real coefficients; the operator V_0 is supposed to be hyperbolic.
(b) Maximums of moduli of coefficient of operators V_s increase not so rapidly as some geometrical progression. We shall use the following notation:

$$W(\lambda, \sigma) = \sum_{s=0}^{\infty} \left[\widetilde{\phi}(\lambda) \right]^{s} V_{s}(\lambda, \sigma), \qquad (1.1.36)$$

$$V(\lambda, \sigma) = V_{0}(\lambda, \sigma) + \widetilde{\phi}(\lambda) V_{1}(\lambda, \sigma) \qquad (1.1.37)$$

Here

$$\widetilde{\phi}(\lambda) = \int_{-\infty}^{+\infty} \phi(t) e^{-i\lambda t} dt = F_{t \to \lambda} \phi$$

is the Fourier–Laplace transform of the function $\phi(t)$; $\lambda = \mu - ip$; $p > 0$..

Expression (1.1.36) and (1.1.37) are called *symbol of the corresponding operators* W and V (for the classes of operators under consideration, such a definition of a symbol coincides with the usual definition of a symbol with accuracy to multiplication by i^{m}). Equations

$$W(\lambda, \sigma) = 0 \qquad (1.1.38)$$

$$V(\lambda, \sigma) = 0 \qquad (1.1.39)$$

will be called the *characteristic* equations for the *corresponding operators*.

Later on we shall see that operator V hyperbolicity in S', stable with respect to small (real) perturbances of coefficients of operators V_{0} and V_{1}, yields operator W hyperbolicity in S', stable with respect to small (real) perturbances of coefficients of all operators V_{s}. Thus the operator V can be considered the principal part of the operator W. From our results, it will also follow that the operator V_{0}, generally speaking, cannot be considered the principal part of the operator W. Below (see Sect. 2.1), we shall begin to study operators with memory with operators of Eq. (1.1.39) type, for which it is possible to obtain the most complete results. Then, we shall pass to more general operators of Eq. (1.1.38) type (see Sects. 2.2, 2.3, 2.4, and 2.5).

1.2 Homogeneous Hyperbolic Polynomials: The Propagation Cone and the Influence Cone

(a) Let $\lambda \in \mathbb{R}^{1}$, $\sigma \in \mathbb{R}^{n}$. We would remind to the reader that a homogeneous polynomial $P(\lambda, \sigma)$ of degree m is called *hyperbolic* (or, to be more precise, *hyperbolic with respect to the variable* λ), if for each σ, the equation $P(\lambda, \sigma) = 0$ has m real roots $\lambda_{1}(\sigma), \ldots, \lambda_{m}(\sigma)$.

In Sect. 1.1.8, we have supposed V_{0} to be a hyperbolic homogeneous operator of m-th order. But it is well known [7] that a homogeneous differential operator is hyperbolic (in D') if and only if its symbol is a hyperbolic polynomial. Hence, in what follows $V_0(\lambda, \sigma)$ is a hyperbolic polynomial.

Furthermore, we recall that a hyperbolic polynomial $V_0(\lambda, \sigma)$ is said to be strictly hyperbolic if for $\sigma \neq 0$ all the roots $\lambda_1(\sigma), \ldots, \lambda_m(\sigma)$ of the equation $V_0(\lambda, \sigma) = 0$ are different.

(b) The set

$$\{\lambda, \sigma | V_0(\lambda, \sigma) = 0\} \subset \mathbb{R}^1 \times \mathbb{R}^n \qquad (1.2.1)$$

is called the *normal cone* for the operator V_0. The connected component of $\mathbb{R}^n \backslash \{\lambda, \sigma | V_0(\lambda, \sigma) = 0\}$ containing the semi-axis $\lambda > 0$ is called the *core of the normal cone* for the operator V_0 and will be denoted by $\circ N$. It is well known that the core of the normal cone for a hyperbolic operator is convex [6, 7].

(c) V_0 is said to be an *operator with bounded normal surface* if the intersection of the normal cone (1.2.1) with the hyperplane $\lambda = 0$ consists of the origin of the coordinates $\lambda = 0$ and $\sigma = 0$. Otherwise, we say that V_0 is an operator with unbounded normal surface (see [1.8]).

(d) Let $g : \lambda, \sigma \to \lambda', \sigma'$ be a linear homogeneous mapping of $\mathbb{R}^1 \times \mathbb{R}^n$ into itself, such that the semi-axis $\lambda' > 0$ belongs to the core of the normal cone.

for V_0. Then the polynomial $V_0(\lambda(\lambda', \sigma'), \sigma(\lambda', \sigma'))$ is hyperbolic with respect to λ'. If the polynomial $V_0(\lambda, \sigma)$ is strictly hyperbolic (with respect to λ), then the polynomial $V_0(\lambda(\lambda', \sigma'), \sigma(\lambda', \sigma'))$ is also strictly hyperbolic (with respect to λ') [6, 7].

(e) Let $t \in \mathbb{R}^1$, $x \in \mathbb{R}^n$. Consider the cone $\circ K$ dual to $\circ N$:

$$\circ K = \{t, x | tx + x \cdot \sigma \geq 0, \text{for} (\lambda, \sigma) \in \circ N\} \subset \mathbb{R}^1 \times \mathbb{R}^n \qquad (1.2.2)$$

The cone $\circ K$ is called the *propagation cone* for the operator V_0.

Let $E_{V_0}(t, x)$ be the operator V_0 fundamental solution describing finite speed wave propagation. Then $\circ K$ coincides with the closure of the convex hull of $\operatorname{supp} E_{V_0}(t, x)$ [6, 7]. The cone $\circ K$ can also be defined as the intersection of the half-space $t \geq 0$ with the closure of the convex hull of the characteristic cone for V_0).

(f) Now, let K be the closure of the convex hull of the set

$$\{t, x | t > 0, x = 0\} \cup \circ K.$$

It is clear that

$$K \supseteq \circ K \qquad (1.2.3)$$

Evidently, $K = \circ K$ if and only if the semi-axis $t > 0$ belongs to the cone $\circ K$. We shall call K the *influence cone* for the operator V_0. It is clear that both K and $\circ K$ are proper cones of the half-space $t > 0$, that is, some cone of the $\{t, x | t \geq \text{const}|x|\}$, const > 0 type contains both K and $\circ K$. We shall denote the bounds of K and $\circ K$ by ∂K and $\partial \circ K$, respectively. Evidently, the intersection $\partial K \cap \partial \circ K$ contains at least one half-line.

(g) Let us define a cone

$$N = \{\lambda, \sigma \,|\, t\lambda + x \cdot \sigma > 0 \,\text{for}\,(t, x) \in K\} \subset \mathbb{R}^1 \times \mathbb{R}^n \qquad (1.2.4)$$

From (1.2.3), it evidently follows that

$$N \subseteq \circ N \qquad (1.2.5)$$

Moreover, $N = \circ N$ if and only if $K = \circ K$. It is easy to see that

$$N = \circ N \cap \{\lambda, \sigma \,|\, \lambda > 0\} \qquad (1.2.6)$$

(h) Consider an example

$$V_0 = \left(\frac{\partial}{\partial t} + \frac{\partial}{\partial x}\right)\left(\frac{\partial}{\partial t} + 2\frac{\partial}{\partial x}\right).$$

Then the cones $\circ N$, $\circ K$, N, K are shown in Fig. 1.1.

(i) It can be shown that in case of $n > 1$, the boundedness of the normal surface for the operator V_0 yields $N = \circ N$ (whence $K = \circ K$).

1.3   The Paley–Wiener-Type Theorems

The definition of the Fourier–Laplace transform for distributions can be found, for example, in [6, 7]. The symbol $F_{y \to z}$ is called the *Fourier–Laplace transform* with respect to variables indicated in the subscript. In case where the distribution $f(y)$, $y \in \mathbb{R}^n$, coincides with a usual locally integrable function, the Fourier–Laplace transform reduces to the integral

$$F_{y \to z} f = \int\limits_{-\infty}^{\infty} \cdots \int\limits_{-\infty}^{\infty} f(y) e^{-izy} \, dy_1 \cdots dy_n \qquad (1.3.1)$$

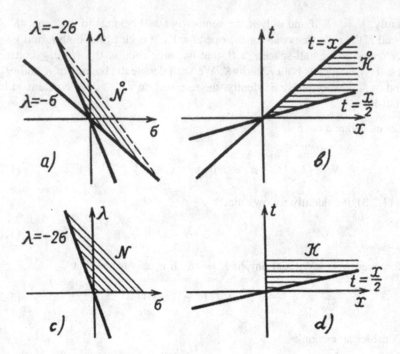

Fig. 1.1 Cones of propagation and of influence (as well as cones dual to them) for the operator from the example (h)

where $zy = z_1 y_1 + \cdots + z_n y_n$. The integral (1.3.1) is supposed to be convergent when Im z belongs to some open set in Im $\mathbb{C}^n$.

Theorem 1.3.1 [9] Let $t \in \mathbb{R}^1$ and let $f(t)$ be a distribution equal to zero for $t < T$ and satisfying the condition $e^{-Mt} f(t) \in S'$ for some $M \in \mathbb{R}^1$. Then the function $F_{t \to \lambda} f$ is holomorphic in the half-plane Im$\lambda < -M$ and satisfies the following estimate

$$| F_{t \to \lambda} f | \le C (1 + |\lambda|)^{\nu} (1 + |M + \mathrm{Im}\lambda|^{\nu}) e^{T\mathrm{Im}\lambda} \tag{1.3.2}$$

for Im $\lambda < -M$. Here $C = \mathrm{const} > 0$, $\nu = \mathrm{const} > 0$.

Conversely, let for Im $\lambda < -M$ the function $g(\lambda)$ be holomorphic and satisfying (1.3.2). Then $g(\lambda)$ is the Fourier–Laplace transform of some distribution equal to zero for $t < T$ and such that $e^{-Mt} f(t) \in S'$.

Theorem 1.3.2 [7, 10] Let $t \in \mathbb{R}^1$, $x \in \mathbb{R}^n$ and suppose K is a closed convex cone in $\mathbb{R}^1 \times \mathbb{R}^n$ satisfying the condition $K \subseteq \{t, x | t \ge \mathrm{const} | x | \}$, const > 0. Furthermore, let $f(t, x)$ be a distribution such that supp $f(t, x) \subseteq K$ and $e^{-Mt} f(t, x) \in S'$ for some $M \in \mathbb{R}^1$. Then the function $F_{t, x \to \lambda, \sigma} f$ is holomorphic in the complex domain

$$\{\lambda, \sigma | t(M + \mathrm{Im}\lambda) + x \cdot \mathrm{Im}\sigma < 0 \, for \, (t, x) \in K\} \tag{1.3.3}$$

$(\lambda \in \mathbb{C}^1, \sigma \in \mathbb{C}^n)$ and satisfies the estimate

$$\begin{aligned}
| F_{t,x \to \lambda, \sigma} f | &\leq C(M + \mathrm{Im}\lambda, \mathrm{Im}\sigma) \cdot (1 + |\lambda| + |\sigma|)^\nu \\
&\times [1 + (|M + \mathrm{Im}\lambda| + |\mathrm{Im}\sigma|)^{-\nu}]
\end{aligned} \tag{1.3.4}$$

in the domain (1.3.3). Here $C(\mathrm{Im}\lambda, \mathrm{Im}\sigma)$ is a function locally bounded on the open set $\{\mathrm{Im}\lambda, \mathrm{Im}\sigma | t \, \mathrm{Im} \, \lambda + x \cdot \mathrm{Im} \, \sigma < 0 \, for \, (t, x) \in K\}$ and satisfying the condition $C(s \, \mathrm{Im} \, \lambda, s \, \mathrm{Im} \, \sigma) = C(\mathrm{Im}\lambda, \mathrm{Im}\sigma)$, for $s > 0$;

$$\nu = \mathrm{const} > 0.$$

Conversely, suppose a function $g(\lambda, \sigma)$ is holomorphic in the domain (1.3.3) and satisfies the estimate (1.3.4) in the mentioned domain. Then $g(\lambda, \sigma)$ is the Fourier–Laplace transform of some distribution $f(t, x)$ such that $\mathrm{supp} \, f(t, x) \subseteq K$ and $e^{-Mt} f(t, x) \in S'$.

Theorem 1.3.3 [6, 11] Let $x \in \mathbb{R}^n$ and suppose $f(x) \in D'$ is a distribution such that $\mathrm{supp} \, f(x) \subseteq Q$, where Q is a bounded open convex set. Then $F_{x \to \sigma} f$ is an entire function satisfying the estimate

$$| F_{x \to \sigma} f | \leq C(1 + |\sigma|)^\nu e^{h(\mathrm{Im}\sigma)} \tag{1.3.5}$$

where

$$C = \mathrm{const} > 0, \quad \nu = \mathrm{const} > 0; h(\eta) = \sup_{x \in Q} (x \cdot \eta), \quad \eta \in \mathbb{R}^n$$

Conversely, suppose an entire function $g(\sigma)$ satisfies the estimate (1.3.5). Then $g(\sigma)$ is the Fourier–Laplace transform of some distribution $f(x) \in D'$ such that $\mathrm{supp} \, f(x) \subseteq Q$.

1.4 A Lemma About Cos-Fourier Transform

Lemma 1.4.1 [12] Let a function $\phi(t)$ be continuous and concave for $t > 0$, locally integrable on $[0, \infty]$ and tending to zero for $t \to +\infty$. Then for $\mu \neq 0$

$$\int_0^\infty \phi(t) \cos \mu t \, dt \geq 0. \tag{1.4.1}$$

Note The improper integral in (1.4.1) should be understood as

$$\lim_{A\to\infty} \int_0^A \phi(t)\cos\mu t\,dt.$$

Proof of the Lemma From conditions of the lemma, it easily follows that $\phi(t)$ is monotone decreasing for $t > 0$. Therefore, the existence of the integral (1.4.1) is obvious. Furthermore, we rewrite the integral (1.4.1) in the form

$$\sum_{n=0}^{\infty} \int_0^{\pi/2} \left\{ \left[\phi\left(\frac{t+2n\pi}{|\mu|}\right) - \phi\left(\frac{\pi-t+2n\pi}{|\mu|}\right) \right] \right.$$
$$\left. - \left[\phi\left(\frac{\pi+t+2n\pi}{|\mu|}\right) - \phi\left(\frac{2\pi-t+2n\pi}{|\mu|}\right) \right] \right\} \frac{\cos t}{|\mu|}\,dt \qquad (1.4.2)$$

Since $\varphi(t)$ is concave, we have

$$\varphi\left(\frac{t+2n\pi}{|\mu|}\right) - \varphi\left(\frac{\pi-t+2\pi n}{|\mu|}\right) \geq \varphi\left(\frac{\pi+t+2n\pi}{|\mu|}\right) - \varphi\left(\frac{2\pi-t+2n\pi}{|\mu|}\right)$$

Thus the expression in braces in (1.4.2) is non-negative, whence follows the required result.

1.5　Lemmas About the Fourier–Laplace Transform of the Function of Memory

Before proceeding further, we recall that everywhere in this chapter we supposed the function of memory $\phi(t)$ to possess properties 1–5 given in Sect. 1.1.8. As before, we denote the Fourier–Laplace transform

$$F_{t\to\lambda}\phi \text{ by } \widetilde{\phi}(\lambda), \lambda = \mu - ip$$

Lemma 1.5.1 Prove that

$$\widetilde{\phi}(\lambda) \to 0 \quad \text{as } p \to +\infty \qquad (1.5.1)$$

uniformly with respect to μ.

Proof Let us represent $\phi(t)$ as a sum

$$\phi_1(t) + [\phi(t) - \phi_1(t)]$$

where the function $\phi_1(t)$, equals zero for $t < 0$ and for $t > 0$ large enough, is continuous for $t > 0$, coincides with $\phi(t)$ for small $t > 0$, and satisfies the condition

$$\int_0^\infty |\phi_1(t)| \, dt \le \varepsilon; \quad \varepsilon > 0 \tag{1.5.2}$$

Note that, first of all, Eq. (1.5.2) yields the inequality

$$|\widetilde{\phi}_1(\lambda)| \le \int_0^\infty |\phi_1(t)| \, e^{-pt} dt \le \varepsilon; p > 0 \tag{1.5.3}$$

Furthermore, one can easily see that under the above assumptions, the following estimate is true

$$|\phi(t) - \phi_1(t)| \le C$$

where C is some constant depending on the choice of ϕ_1. Therefore

$$|\widetilde{\phi}(\lambda) - \widetilde{\phi}_1(\lambda)| \le \int_0^\infty |\phi(t) - \phi_1(t)| \, e^{-pt} \, dt \le C \int_0^\infty e^{-pt} \, dt = \frac{C}{p}; \quad p$$

$$> 0 \tag{1.5.4}$$

Thus (1.5.3 and 1.5.4) yield

$$|\widetilde{\phi}(\lambda)| \le |\widetilde{\phi}_1(\lambda)| + |\widetilde{\phi}(\lambda) - \widetilde{\phi}_1(\lambda)| \le \varepsilon + \frac{C}{p}; \quad p > 0$$

Since $\varepsilon > 0$ can be taken however small, the required result follows.

Lemma 1.5.2 Let $p > 0$. Then

$$\mu \int_0^\infty \phi(t) e^{-pt} \sin \mu t \, dt \ge \begin{cases} \phi\left(\dfrac{\pi}{2|\mu|}\right) \exp\left(\dfrac{-p\pi}{2|\mu|}\right), & \text{for } \mu \ne 0 \\ 0, & \text{for } \mu = 0 \end{cases} \tag{1.5.5}$$

Proof To be specific, suppose $\mu > 0$. We have

$$\mu \int_0^\infty \phi(t)\,e^{-pt}\,\sin\mu t\,dt = \mu \int_0^{\pi/(2\mu)} \phi(t)\,e^{-pt}\,\sin\mu t\,dt$$

$$+\mu \int_0^\infty \phi\left(t+\frac{\pi}{2\mu}\right) e^{-p\left(t+\frac{\pi}{2\mu}\right)} \cos\mu t\,dt \qquad (1.5.6)$$

From Properties 1–5 of the function $\phi(t)$, it follows that the function

$$\varphi(t) = \phi\left(t+\frac{\pi}{2\mu}\right) e^{-p\left(t+\frac{\pi}{2\mu}\right)}$$

satisfies the conditions of Lemma 1.4.1. Therefore, by virtue of Lemma 1.4.1, the second summand on the right-hand side of (1.5.6) is non negative. Thus

$$\mu \int_0^\infty \phi(t)\,e^{-pt}\,\sin\mu t\,dt \geq \mu \int_0^{\pi/(2\mu)} \phi(t)\,e^{-pt}\,\sin\mu t\,dt$$

$$= \int_0^{\pi/2} \phi\left(\frac{t}{\mu}\right) e^{-pt/\mu}\,\sin t\,dt$$

$$\geq \phi\left(\frac{\pi}{2\mu}\right) e^{-p\pi/(2\mu)} \int_0^{\pi/2} \sin\mu t\,dt,$$

which gives the result required.

Lemma 1.5.3 Let $p > 0$, $|\mu| \geq M$, where $M > 0$ is sufficiently large. Then

$$0 \leq \mu \int_0^\infty \phi(t)\,e^{-pt}\,\sin\mu t\,dt \leq \pi^2 \phi\left(\frac{\pi}{|\mu|}\right) \qquad (1.5.7)$$

Proof To be specific, suppose $\mu > 0$. Then by virtue of the decrease of the function $\phi(t)$, for $t > 0$

$$0 \le \mu \int\limits_0^\infty \phi(t)\, e^{-pt} \sin\mu t\, dt \le \mu \int\limits_0^{\pi/\mu} \phi(t)\, e^{-pt} \sin\mu t\, dt$$

$$\le \mu \int\limits_0^{\pi/\mu} \phi(t)\mu t\, dt$$

By virtue of fifth property of $\phi(t)$, the function $t\phi(t)$ is increasing for small $t > 0$. Therefore, provided μ is sufficiently large, we have

$$\mu \int\limits_0^{\pi/\mu} \phi(t)\mu t\, dt = \mu^2 \int\limits_0^{\pi/\mu} t\phi(t)\, dt \le \mu^2 \left[\frac{\pi}{\mu}\phi\left(\frac{\pi}{\mu}\right) \right] \frac{\pi}{\mu} = \pi^2 \phi\left(\frac{\pi}{\mu}\right),$$

which proves the lemma.

Lemma 1.5.4 Let $p > 0$. Then

$$\mathrm{Im}\left[\lambda\widetilde{\phi}(\lambda)\right] \le \begin{cases} -\phi\left(\dfrac{\pi}{2\,|\,\mu\,|}\right) \exp\left(\dfrac{-p\pi}{2\,|\,\mu\,|}\right), & \text{for } \mu \ne 0 \\ \\ 0, & \text{for } \mu = 0 \end{cases} \qquad (1.5.8)$$

Proof Using Lemma 1.4.1, we have

$$\mathrm{Im}\left[\lambda\widetilde{\phi}(\lambda)\right] = -p \int\limits_0^\infty \phi(t) e^{-pt} \cos\mu t\, dt - \mu \int\limits_0^\infty \phi(t) e^{-pt} \sin\mu t\, dt$$

$$\le -\mu \int\limits_0^\infty \phi(t) e^{-pt} \sin\mu t\, dt$$

$$(1.5.9)$$

Now, the required result follows from Lemma 1.5.2.

Lemma 1.5.5 Let $p \ge M$, where $M > 0$ is large enough. Then the following estimate holds

$$\left| \mathrm{Re}\left[\lambda\widetilde{\phi}(\lambda)\right] \right| \le \mathrm{const}\left| \mathrm{Im}\left[\lambda\widetilde{\phi}(\lambda)\right] \right|, \qquad \mathrm{const} > 0 \qquad (1.5.10)$$

Proof We have

$$
\lambda\widetilde{\phi}(\lambda) = \left(\mu \int\limits_0^\infty \phi(t)\, e^{-pt} \cos \mu t\, dt - p \int\limits_0^\infty \phi(t)\, e^{-pt} \sin \mu t\, dt \right)
$$
$$
+ i \left(-\mu \int\limits_0^\infty \phi(t)\, e^{-pt} \sin \mu t\, dt - p \int\limits_0^\infty \phi(t)\, e^{-pt} \cos \mu t\, dt \right)
\tag{1.5.11}
$$

Clearly, both summands inside of the extreme right parentheses in (1.5.11) are negative for all μ. To be specific, we restrict ourselves to the case of $\mu > 0$.

At first, we suppose

$$
\frac{\mu}{p} \geq \nu > 0, \quad p \geq M > 0
$$

Then the first summand of the imaginary part of (1.5.11), after being multiplied by an appropriate constant, dominates the second summand of the real part of (1.5.11).

$$
\mathrm{const}\left| \mu \int\limits_0^\infty \phi(t) e^{-pt} \sin \mu t\, dt \right| \geq \left| p \int\limits_0^\infty \phi(t)\, e^{-pt} \sin \mu t\, dt \right|;
$$

where $\mathrm{const} = \mathrm{const}\,(\nu) > 0$.

Let us demonstrate that the first summand of the imaginary part of (1.5.11) dominates also the first summand of the real part of (1.5.11). For p and μ satisfying the above conditions Lemma 1.5.2 yields the following inequality

$$
\int\limits_0^\infty \phi(t)\, e^{-pt} \sin \mu t\, dt \geq \frac{\mathrm{const}}{\mu}\, \phi\left(\frac{\pi}{2\mu}\right); \quad \mathrm{const} = \mathrm{const}\,(\nu) > 0
\tag{1.5.12}
$$

Furthermore, without loss of generality, we can consider $\mu > 0$ sufficiently large (since $M > 0$ can be taken large enough). Now, taking into account both the decrease of $\phi(t)$ and its fifth property,

$$0 \le \int_0^\infty \phi(t)\, e^{-pt} \cos \mu t \, dt = \frac{1}{\mu} \int_0^{\pi/2} \phi\left(\frac{t}{\mu}\right) e^{-pt/\mu} \cos t \, dt$$

$$+ \frac{1}{\mu} \int_{\pi/2}^\infty \phi\left(\frac{t}{\mu}\right) e^{-pt/\mu} \cos t \, dt$$

$$\le \frac{1}{\mu} \int_0^{\pi/2} \phi\left(\frac{t}{\mu}\right) e^{-pt/\mu} \, dt$$

$$- \frac{1}{\mu} \int_0^\infty \phi\left(\frac{t + \frac{\pi}{2}}{\mu}\right) \exp\left[-\frac{p\left(t + \frac{\pi}{2}\right)}{\mu}\right] \sin t \, dt \qquad (1.5.13)$$

$$\le \frac{1}{\mu} \int_0^{\pi/2} \phi\left(\frac{t}{\mu}\right) e^{-pt/\mu} \, dt$$

$$= \frac{1}{\mu} \int_0^{\pi/2} \phi\left(\frac{t}{\mu}\right)\left(\frac{t}{\mu}\right)^\gamma \left(\frac{t}{\mu}\right)^{-\gamma} e^{-pt/\mu} \, dt$$

$$\le \frac{1}{\mu} \phi\left(\frac{\pi}{2\mu}\right)\left(\frac{\pi}{2\mu}\right)^\gamma \int_0^{\pi/2} \left(\frac{t}{\mu}\right)^{-\gamma} e^{-t/\nu} \, dt$$

$$\le \frac{\mathrm{const}\,(\nu)}{\mu} \phi\left(\frac{\pi}{2\mu}\right); \quad \mathrm{const}\,(\nu) > 0$$

Formulas (1.5.12 and 1.5.13) yield

$$\mathrm{const}_1(\nu)\left| \mu \int_0^\infty \phi(t)\, e^{-pt} \sin \mu t \, dt \right| \ge \left| \mu \int_0^\infty \phi(t)\, e^{-pt} \cos \mu t \, dt \right|;$$

where $\mathrm{const}_1(\nu) > 0$. Thus for $\mu/p \ge \nu > 0$; $p \ge M > 0$, the required result is proved.

Now let $0 \le \mu/p \le \nu$, $p \ge M > 0$. Then the second summand of the imaginary part of (1.5.11), after being multiplied by a constant, dominates the first summand of the real part of (1.5.11). Let us demonstrate that it dominates also the second summand of the real part of (1.5.11). Making use of concavity of the function

$$-[\phi(t)\, e^{-pt}]'_t, \quad \text{for } t > 0$$

from Property 3 of the function $\phi(t)$, and taking into account Lemma 1.5.1, we have

$$\int\limits_0^\infty \phi(t)\,e^{-pt}\cos\mu t\,dt = -\frac{1}{\mu}\int\limits_0^\infty [\phi(t)\,e^{-pt}]'_t\,\sin\mu t\,dt$$

$$\geq -\frac{1}{\mu}\int\limits_0^{\pi/(2\mu)} [\phi(t)e^{-pt}]'_t\,\sin\mu t\,dt$$

$$\geq -\frac{2}{\pi\mu}\int\limits_0^{\pi/(2\mu)} \mu t\,d[\phi(t)e^{-pt}] \tag{1.5.14}$$

$$= -\frac{1}{\mu}\phi\left(\frac{\pi}{2\mu}\right)\exp\left(-\frac{p\pi}{2\mu}\right)$$

$$+\frac{2}{\pi}\int\limits_0^{\pi/2\mu}\phi(t)\,e^{-pt}\,dt$$

On the other hand, from the decrease of $\phi(t)$, for $t > 0$, it follows that

$$0 \leq \int\limits_0^\infty \phi(t)\,e^{-pt}\sin\mu t\,dt \leq \int\limits_0^{\pi/\mu} \phi(t)\,e^{-pt}\,dt \leq 2\int\limits_0^{\pi/(2\mu)}\phi(t)\,e^{-pt}\,dt \tag{1.5.15}$$

We want to demonstrate that the quantity on the extreme right-hand side of (1.5.14) dominates the quantity on the extreme right-hand side of (1.5.15). In other words, we want to find a small constant $\kappa > 0$ such that

$$-\frac{1}{\mu}\phi\left(\frac{\pi}{2\mu}\right)\exp\left(-\frac{p\pi}{2\mu}\right) + \frac{2}{\pi}\int\limits_0^{\pi/(2\mu)}\phi(t)\,e^{-pt}\,dt \geq 2\kappa\int\limits_0^{\pi/(2\mu)}\phi(t)\,e^{-pt}\,dt \tag{1.5.16}$$

That is

$$\frac{2}{\pi}(1-\pi\kappa)\int\limits_0^{\pi/(2\mu)}\phi(t)\,e^{-pt}\,dt \geq \frac{1}{\mu}\phi\left(\frac{\pi}{2\mu}\right)\exp\left(-\frac{p\pi}{2\mu}\right) \tag{1.5.17}$$

for values of p and μ under consideration. But, by virtue of concavity of the function, $\phi(t)\,e^{-pt}$, for $t > 0$, we have

$$\int\limits_{0}^{\pi/(2\mu)} \phi(t)\, e^{-pt}\, dt \geq \frac{\pi}{2\mu}\, \phi\!\left(\frac{\pi}{4\mu}\right) \exp\!\left(-\frac{p\pi}{4\mu}\right)$$

for each $p > 0$. Therefore, Eq. (1.5.17) will be established if we demonstrate the validity of the inequality

$$(1 - \pi\kappa) \exp\!\left(-\frac{p\pi}{4\mu}\right) \phi\!\left(\frac{\pi}{4\mu}\right) \geq \exp\!\left(-\frac{p\pi}{2\mu}\right) \phi\!\left(\frac{\pi}{2\mu}\right)$$

which, in its turn, can be deduced from the inequality

$$(1 - \pi\kappa) \exp\!\left(\frac{\pi}{4\nu}\right) \phi\!\left(\frac{\pi}{4\mu}\right) \geq \phi\!\left(\frac{\pi}{2\mu}\right)$$

However, for $\kappa > 0$ small enough, the last inequality follows from the decrease of $\phi(t)$ for $t > 0$. The lemma is proved.

Lemma 1.5.6 For $p = \text{const} > 0$, $|\mu| \to \infty$

$$\left|\tilde{\phi}(\lambda)\right| \leq \frac{\text{const}}{|\mu|}\, \phi\!\left(\frac{\pi}{2|\mu|}\right) \tag{1.5.18}$$

Proof
From formulas (1.5.7 and 1.5.13)

Lemma 1.5.7 For

$$\mu/p \geq \nu > 0,\, p \geq M > 0$$
$$-\frac{\pi}{2} \leq \arg\tilde{\phi}(\lambda) \leq -\theta_0 < 0; \tag{1.5.19}$$

$$\mu/p \leq -\nu < 0,\, p \geq M > 0$$
$$\frac{\pi}{2} \geq \arg\tilde{\phi}(\lambda) \geq \theta_0 > 0 \tag{1.5.20}$$

Here M is supposed to be sufficiently large.

Proof Let us prove, for example, the proposition (a) of the lemma. From the proof of Lemma 1.5.5, it follows that for $\mu/p \geq \nu$, $p \geq M$, the inequality

$$0 \leq \int\limits_{0}^{\infty} \phi(t)\, e^{-pt} \cos \mu t\, dt \leq \text{const} \int\limits_{0}^{\infty} \phi(t)\, e^{-pt} \sin \mu t\, dt \tag{1.5.21}$$

holds for $\text{const} = \text{const}(\nu) > 0$. That is

$$0 \leq \operatorname{Re} \widetilde{\phi}(\lambda) \leq -\text{const}\left[\operatorname{Im}\widetilde{\phi}(\lambda)\right],$$

which gives the required result.

Lemma 1.5.8

$$\operatorname{Im}\left[\lambda\widetilde{\phi}(\lambda)\right] \to -\infty, \quad \text{as } p \to +\infty \tag{1.5.22}$$

uniformly with respect to μ.

Proof To be specific, we consider $\mu > 0$. At first, let $\mu/p \geq \nu > 0, p \geq M > 0$. Then (1.5.9 and 1.5.12) yield

$$\operatorname{Im}\left[\lambda\widetilde{\phi}(\lambda)\right] \leq -\mu \int\limits_0^\infty \phi(t)\, e^{-pt} \sin \mu t\, dt \leq -\text{const}\, \phi\left(\frac{\pi}{2\mu}\right)$$
$$\leq -\text{const}\, \phi\left(\frac{\pi}{2\nu p}\right) \to -\infty, \quad \text{as } p \to +\infty; \tag{1.5.23}$$

for const $=$ const $(\nu) > 0$..

Now, let $\frac{\mu}{p} \leq \nu$, $p > M > 0$. Then (1.5.5) yields

$$\operatorname{Im}\left[\lambda\widetilde{\phi}(\lambda)\right] \leq -p \int\limits_0^\infty \phi(t) e^{-pt} \cos \mu t\, dt \tag{1.5.24}$$

Furthermore, it follows from the proof of Lemma 1.5.5 that, for p and μ under consideration, the inequality

$$\int\limits_0^\infty \phi(t)e^{-pt} \cos \mu t\, dt \geq \text{const} \int\limits_0^{\pi/(2\mu)} \phi(t)e^{-pt}\, dt; \tag{1.5.25}$$

holds for const $=$ const (ν). See (1.5.14 and 1.5.16)).

Now, (1.5.24 and 1.5.25) give

$$\operatorname{Im}\left[\lambda\widetilde{\phi}(\lambda)\right] \leq -\text{const}\, p \int\limits_0^{\pi/(2\mu)} \phi(t)e^{-pt}\, dt = -\text{const} \int\limits_0^{\pi p/(2\mu)} \phi\left(\frac{t}{p}\right)e^{-t}\, dt \tag{1.5.26}$$

$$\leq -\text{const} \int\limits_{0}^{\pi/(2\nu)} \phi\left(\frac{t}{p}\right) e^{-t} dt \to -\infty, \quad \text{as } p \to +\infty$$

From (1.5.23 and 1.5.26), the required result follows.

Lemma 1.5.9 Let

$$\phi(t) \leq \text{const}\left(\ln\frac{1}{t}\right), \quad \text{const} > 0, \tag{1.5.27}$$

for small $t > 0$. Then

$$\left|\lambda\widetilde{\phi}(\lambda)\right| \leq \text{const}\left[\ln\left(|\lambda| + 1\right) + 1\right], \quad \text{const} > 0 \tag{1.5.28}$$

Proof We have

$$-\text{Im}\left[\lambda\widetilde{\phi}(\lambda)\right] = \mu \int\limits_{0}^{\infty} \phi(t) e^{-pt} \sin\mu t + p \int\limits_{0}^{\infty} \phi(t) e^{-pt} \cos\mu t \, dt$$

$$\equiv I_1 + I_2$$

Consider at first the summand

$$I_1 = \mu \int\limits_{0}^{\infty} \phi(t) e^{-pt} \sin\mu t \, dt$$

For $|\mu|$ large enough, we have by virtue of Lemma 1.5.3:

$$0 \leq I_1 \leq \pi^2 \phi\left(\frac{\pi}{|\mu|}\right) \leq \text{const}\left[\ln\left(|\mu| + 1\right)\right], \quad \text{const} > 0$$

Now, let $|\mu| \leq$ const. Then evidently $0 \leq I_1 \leq$ const. Therefore

$$0 \leq I_1 \leq \text{const}\left[\ln\left(|\mu| + 1\right) + 1\right], \quad \text{const} > 0, \tag{1.5.29}$$

for all μ.

Now consider the summand

$$I_2 = p \int_0^\infty \phi(t) e^{-pt} \cos \mu t \, dt$$

For all μ, we have

$$0 \le I_2 \le p \int_0^\infty \phi(t) e^{-pt} \, dt$$

By virtue of the conditions of the lemma, one can construct a function $\phi_1(t)$, which coincides with $\phi(t)$, for small $t > 0$, satisfies the inequality

$$\phi_1(t) \le \alpha \ln \frac{1}{t}, \quad \alpha > 0, \tag{1.5.30}$$

on the whole of semi-axis $t > 0$, coincides with $\beta \ln (1/t)$, $\beta > 0$, for large $t > 0$, and is continuous for $t > 0$. Then, on the one hand, (1.5.30) yields

$$p \int_0^\infty \phi_1(t) e^{-pt} \, dt \le \alpha p \int_0^\infty \left(\ln \frac{1}{t} \right) e^{-pt} \, dt = \alpha(C + \ln p) \tag{1.5.31}$$

where $C = 0.57...$ is the *Euler's constant*. On the other hand

$$p \int_0^\infty [\phi(t) - \phi_1(t)] e^{-pt} \, dt = p \int_0^\infty [\phi(t) - \phi_1(t)] e^{-\varepsilon t} e^{-(p-\varepsilon)t} \, dt, \tag{1.5.32}$$

for $\varepsilon > 0$. By virtue of the above assumptions, it is evident that

$$|[\phi(t) - \phi_1(t)] e^{-\varepsilon t}| \le \text{const}, \quad t \ge 0.$$

Therefore, it follows from (1.5.32) that

$$\left| p \int_0^\infty [\phi(t) - \phi_1(t)] e^{-pt} \, dt \right| \le \text{const} \left[p \int_0^\infty e^{-(p-\varepsilon)t} \, dt \right] \tag{1.5.33}$$

$$\le \text{const} \frac{p}{p - \varepsilon} \le \text{const for} \, p \ge M > \varepsilon$$

Thus, on account of (1.5.31 and 1.5.33), we have for $p \ge M$,

$$0 \leq I_2 \leq p \int\limits_0^\infty \phi(t)e^{-pt}\,\mathrm{d}t$$

$$\leq \left| p \int\limits_0^\infty \phi_1(t)e^{-pt}\mathrm{d}t \right| + \left| p \int\limits_0^\infty [\phi(t) - \phi_1(t)]e^{-pt}\mathrm{d}t \right| \tag{1.5.34}$$

$$+ \left| p \int\limits_0^\infty [\phi(t) - \phi_1(t)]e^{-pt}\mathrm{d}t \right|,$$

$$\leq \mathrm{const}[\ln(p+1) + 1]$$

for const > 0.

Now, (1.5.29 and 1.5.34) and Lemma 1.5.5 give the result required.

Lemma 1.5.10 Let

$$\phi(t) \leq \mathrm{const}(t^{-\alpha}), \quad 0 < \alpha < 1, \quad \mathrm{const} > 0 \tag{1.5.35}$$

for small $t > 0$. Then

$$\left| \lambda \widetilde{\phi}(\lambda) \right| \leq \mathrm{const}\, |\lambda|^\alpha, \quad \mathrm{const} > 0, \tag{1.5.36}$$

for $p \geq M > 0$. The proof of this lemma is similar to the one of Lemma 1.5.9.

References

1. Lokshin, A.A., Suvorova, J.V.: *Mathematical Theory of Wave Propagation in Media with Memory*, pp. 1–151. Moskow University Press, Moscow (1982)
2. Kelbert, M. J. and Chaban, I. A. (1986). *Izv. Akad. Nauk. SSSR MZhG* 4, 164.
3. Rabotnovy, J.N.: *Elements of Hereditary Mechanics of Solids*, pp. 1–383. Nauka, Moscow (1977)
4. Christensen, R.M.: *Theory of Viscoelasticity*. Academic Press, New York (1971)
5. Aki, K., Richards, P.: *Quantitative Seismology*, vol. 1. WH. Freeman and Company, San Francisco (1980)
6. Hormander, L.: *Linear Partial Differential Operators*. Springer, Berlin (1963)
7. Atiah, M.F., Bott, R., Görding, L.: *Acta Math.* **24**, 109 (1970)
8. John, F.: *Plane Waves and Spherical Means Applied to Partial Differential Equations*. Interscience Publishers, New York (1955)
9. Volevitch, L.R., Gindikin, S.G.: *Usp. Mat. Nauk.* **2**, 65 (1972)
10. Vladimirov, V.S.: *Distributions in Mathematical Physics*, pp. 1–318. Nauka, Moscow (1979)
11. Plancherel, M., Polya, G.: *Comm. Math. Helv.* **9**, 224 (1937)
12. Ahieser, N.I.: *Lectures on Approximation Theory*, pp. 1–407. Nauka, Moscow (1965)

Chapter 2
General Hyperbolic Operators with Memory

This chapter is devoted to the study of conditions of hyperbolicity for intergro-differential operators of (1.1.34) and (1.1.35) types, that is, conditions under which the operators in question describe finite speed wave propagation. In Sects. 2.1 and 2.2, we deal with the one-dimensional case; Sects. 2.3, 2.4, and 2.5 are devoted to the case of n spatial dimensions.

2.1 One-Dimensional Case: The Simplest Hyperbolic Operators with Memory

Let V be an operator with the symbol

$$V(\lambda, \sigma) = V_0(\lambda, \sigma) + \tilde{\phi}(\lambda) V_1(\lambda, \sigma); \lambda \in \mathbb{C}^1, \sigma \in \mathbb{C}^1 \qquad (2.1.1)$$

satisfying the conditions formulated in Sect. 1.1.8 in the one-dimensional case. The symbol of a homogeneous hyperbolic operator V_0 of order m, clearly, can be decomposed into a product of first-order factors

$$V_0(\lambda, \sigma) = V_0(1,0) \prod_{j=1}^{m} (\lambda - c_j \sigma) \qquad (2.1.2)$$

If the operator V_0 is strictly hyperbolic with a bounded normal surface, then all the $c_j, j = 1,\ldots, m$ are different and distinct from zero and hence the coefficients

A. A. Lokshin, *Tauberian Theory of Wave Fronts in Linear Hereditary Elasticity*, https://doi.org/10.1007/978-981-15-8578-4_2

$$k_j \equiv \left. \frac{V_1(\lambda, \sigma)}{\lambda \frac{\partial}{\partial \lambda} V_0(\lambda, \sigma)} \right|_{\lambda = c_j \sigma, \sigma \neq 0} \qquad ; j = 1, \ldots, m \qquad (2.1.3)$$

are defined correctly.

Theorem 2.1.1 Let V_0 be a strictly hyperbolic operator with a bounded normal surface. Then the operator V is hyperbolic in S' if and only if the inequalities

$$k_j \geq 0; \quad j = 1, \ldots, m \qquad (2.1.4)$$

hold. The condition (2.1.4) if satisfied, supp $E(t, x) \subseteq K$ and $E(t, x) \not\equiv 0$ in a however small neighbourhood of an arbitrary point $P \in \partial K \cap \partial \circ K$. (Here $E(t, x)$ is the operator V fundamental solution describing finite speed wave propagation; K and $\circ K$ are the influence cone and the propagation cone for the operator V_0, respectively).

Note About Uniqueness From the uniqueness theorem for equations in convolutional algebras [1], it follows that the operator V has no more than one fundamental solution $E(t, x) \in D'$ with a support contained in a proper cone of the half-space $t \geq 0$. Really, let $E(t, x)$ and $E_1(t, x) \in D'$ be the operator V fundamental solutions vanishing outside a proper cone of the half-space $t \geq 0$. Then

$$V(E(t, x) - E_1(t, x)) = 0$$

We shall denote the convolution with respect to t, x by $\underset{t,x}{*}$. Then

$$E(t, x) - E_1(t, x) = (E(t, x) - E_1(t, x)) \underset{t,x}{*} \delta(t)\delta(x)$$
$$= (E(t, x) - E_1(t, x)) \underset{t,x}{*} VE_1(t, x).$$

Since V is a convolution operator, the previous expression, by commutativity of convolution, equals

$$V(E(t, x) - E_1(t, x)) \underset{t,x}{*} E_1(t, x) = 0 \underset{t,x}{*} E_1(t, x) = 0.$$

Thus $E(t, x) = E_1(t, x)$..

In what follows, the reader must have in mind the above note. Before proceeding, we have to establish the following algebraic result.

Lemma 2.1.1 Let $\sigma_j, j = 1, \ldots, m$, be the roots of the characteristic equation

$$V(\lambda, \sigma) \equiv V_0(\lambda, \sigma) + \tilde{\phi}(\lambda)V_1(\lambda, \sigma) = 0 \qquad (2.1.5)$$

where $V_0(\lambda, \sigma)$ is the symbol of a strictly hyperbolic operator with bounded normal surface. Then for large $- \mathrm{Im}\,\lambda$, the roots $\sigma_j(\lambda)$ are continuous[1] and satisfy the relations:

(a) Uniformly with respect to $\mathrm{Re}\lambda$,

$$\sigma_j(\lambda) = \frac{\lambda}{c_j} + \frac{k_j[1 + o(1)]}{c_j} \lambda \widetilde{\phi}(\lambda); \tag{2.1.6}$$

for $j = 1, \ldots, m$; $\mathrm{Im}\,\lambda \to - \infty$.

(b) Uniformly with respect to $\mathrm{Re}\lambda$,

$$\mathrm{Im}\,\sigma_j(\lambda) = \frac{\mathrm{Im}\lambda}{c_j} + \frac{k_j[1 + o(1)]}{c_j} \mathrm{Im}\left[\lambda\widetilde{\phi}(\lambda)\right]; \tag{2.1.7}$$

for $j = 1, \ldots, m$; $\mathrm{Im}\,\lambda \to - \infty$.

(c) For real p,

$$\mathrm{Im}\,\sigma_j(-ip) = -\frac{p}{c_j}[1 + o(1)]; \tag{2.1.8}$$

for $j = 1, \ldots, m$; $p \to + \infty$.

Proof

(a) Let us write out expansions of the symbols $V_0(\lambda, \sigma)$ and $V_1(\lambda, \sigma)$ in powers of $\lambda - c_1\sigma$. For $a_0 \neq 0$, we have

$$V_0(\lambda, \sigma) = (\lambda - c_1\sigma)\left[a_{m-1}(\lambda - c_1)^{m-1} + a_{m-2}(\lambda - c_1)^{m-2}\sigma + \cdots + a_0\sigma^{m-1}\right],$$
$$V_1(\lambda, \sigma) = b_m(\lambda - c_1\sigma)^m + b_{m-1}(\lambda - c_1\sigma)^{m-1}\sigma + \cdots + b_0\sigma^m$$

Now, the characteristic Eq. (2.1.5) assumes the form

$$(\lambda - c_1\sigma)\left[a_{m-1}(\lambda - c_1\sigma)^{m-1} + a_{m-2}(\lambda - c_1\sigma)^{m-2}\sigma + \cdots + a_0\sigma^{n-1}\right]$$
$$+ \widetilde{\phi}(\lambda)\left[b_m(\lambda - c_1\sigma)^m + b_{m-1}(\lambda - c_1\sigma)^{m-1}\sigma + \cdots + b_0\sigma^m\right] = 0 \tag{2.1.9}$$

Let us introduce a change:

---

[1]Moreover, $\sigma_j(\lambda)$ prove to be holomorphic [2].

$$z = \frac{\lambda - c_1 \sigma}{\sigma \widetilde{\phi}(\lambda)} \qquad (2.1.10)$$

After substituting (2.1.10) into (2.1.9), we obtain by cancellation by $\sigma^m \widetilde{\phi}(\lambda)$:

$$z\left\{a_{m-1}\left[\widetilde{\phi}(\lambda)\right]^{m-1} z^{m-1} + \cdots + a_0\right\} + b_m\left[\widetilde{\phi}(\lambda)\right]^m z^m + \cdots + b_0 = 0. \qquad (2.1.11)$$

Since $\widetilde{\phi}(\lambda) \to 0$ uniformly with respect to $\mathrm{Re}\lambda$ as $\mathrm{Im}\lambda \to -\infty$ (see Lemma 1. 5.1), it follows that the limiting equation for (2.1.11) (as $\mathrm{Im}\lambda \to -\infty$) has the form

$$a_0 z + b_0 = 0, \quad a_0 \neq 0.$$

Now, from the theorem about implicit functions, it easily follows that for $-\mathrm{Im}\,\lambda$ large enough, (2.1.11) has a root $z_1(\lambda)$ which is continuous in λ and can be represented in the form

$$z_1(\lambda) = -\frac{b_0}{a_0} + o(1)$$

Here the quantity $o(1)$ tends to zero uniformly with respect to $\mathrm{Re}\lambda$ as $\mathrm{Im}\lambda \to -\infty$. Furthermore, it is easy to see that

$$\frac{b_0}{a_0} = c_1 k_1$$

where k_1 is defined by (2.1.3) with $j = 1$. Therefore, the characteristic Eq. (2.1.5) has for large $-\mathrm{Im}\,\lambda$ a continuoues root:

$$\sigma_1(\lambda) = \frac{\lambda}{c_1 + z_1(\lambda)\widetilde{\phi}(\lambda)} = \frac{\lambda}{c_1} + \frac{k_1 + o(1)}{c_1} \lambda\widetilde{\phi}(\lambda)$$

It is easy to see that in case of $k_1 = 0$, the equality $\sigma_1(\lambda) = \lambda_1/c$ holds. Therefore, it is possible to rewrite the previous expression for $\sigma_1(\lambda)$ in the form

$$\sigma_1(\lambda) = \frac{\lambda}{c_1} + \frac{k_1[1 + o(1)]}{c_1} \lambda\widetilde{\phi}(\lambda)$$

The rest of the roots of the characteristic equation can be calculated in a similar way.

(b) The assertion of this point of the lemma follows from the assertion of the point (a) by virtue of Lemma 1.5.5.

(c) The assertion of this point follows from the one of the point (b) by virtue of Lemma 1.5.1.

Proof of the Theorem: Necessity   Let, for example, $k_1 < 0$ and suppose the operator V has a fundamental solution $E(t, x)$ describing finite speed wave propagation and satisfying the condition $e^{-M_0 t} E(t,x) \in S'$ for some $M_0 > 0$.

Applying the Fourier–Laplace transform to the equality

$$VE(t,x) = \delta(t)\delta(t)$$

we have

$$
\begin{aligned}
F_{t,x\to\lambda,\sigma} E &= [i^m V(\lambda,\sigma)]^{-1} \\
&= \left\{ i^m \left[V_0(0,1) + \widetilde{\phi}(\lambda) V_1(0,1) \right] \prod_{j=1}^{m} [\sigma - \sigma_j(\lambda)] \right\}^{-1}.
\end{aligned}
\tag{2.1.12}
$$

Our purpose is to obtain a contradiction by finding for the function (2.1.12) singularities of the sort which (2.1.12) cannot have by virtue of Theorem 1.3.2.

Let us study the sign of $c_1 \, \mathrm{Im}\, \sigma_1(\lambda)$ along the straight line $\mathrm{Im}\lambda = -M$ where $M > 0$ is large enough. By virtue of Lemma 2.1.1(c), we have for $\mathrm{Re}\lambda = 0$

$$c_1 \mathrm{Im}\sigma_1(-iM) < 0 \tag{2.1.13}$$

Now, let $\mathrm{Re}\lambda \to +\infty$. Then it follows from Lemma 2.1.1 (b) that

$$c_1 \mathrm{Im}\sigma_1(\lambda) = \mathrm{Im}\lambda + [k_1 + \varepsilon_M(\lambda)]\,\mathrm{Im}\left[\lambda\widetilde{\phi}(\lambda)\right]$$

where one can consider $|\varepsilon_M(\lambda)| < k_1/2$ if M is chosen sufficiently large. Lemma 1.5.4 yields

$$-\mathrm{Im}\left[\lambda\widetilde{\phi}(\lambda)\right] \to +\infty, \text{ for } \mathrm{Im}\lambda = -M,\ \mathrm{Re}\,\lambda \to +\infty$$

Therefore

$$c_1 \mathrm{Im}\sigma_1(\lambda) > 0, \quad \text{for } \mathrm{Im}\lambda = -M,\ \mathrm{Re}\,\lambda \to +\infty \tag{2.1.14}$$

(since $k_1 < 0$). By virtue of continuity of the root $\sigma_1(\lambda)$, it follows from (2.1.13, 2.1.14) that on the straight line $\mathrm{Im}\lambda = -M$ there exists a point λ_0 such that

$$\mathrm{Im}\sigma_1(\lambda_0) = 0. \tag{2.1.15}$$

Now, from (2.1.15), we obtain that the function (2.1.12) has a singularity for $\mathrm{Im}\lambda = -M,\ \sigma \in \mathbb{R}^1$. Since $M > 0$ can be chosen however large, it follows from the above-mentioned that the function (2.1.12) cannot be the Fourier–Laplace transform of a function $E(t, x)$ which describes finite speed wave propagation and satisfies the condition $e^{-M_0 t} E(t, x) \in S'$. Thus we have arrived at the desired contradiction.

Sufficiency The function $[i^n V(\lambda, \sigma)]^{-1}$ being the Fourier–Laplace transform of a distribution $E(t, x)$ such that supp $E(t, x) \subseteq K$ and $e^{-Mt} E(t, x) \in S'$ for some $M > 0$ if demonstrated, then $E(t, x)$ being the desired fundamental solution for the operator V will be also established. By virtue of Theorem 1.3.2, it suffices to establish the inequality

$$\frac{1}{|V(\lambda, \sigma)|} \leq \text{const} \tag{2.1.16}$$

on the set

$$\{\lambda, \sigma \,|\, \text{Im}\lambda \leq \min \left[(1 + \delta)c_1 \,\text{Im}\sigma, \ldots, (1 + \delta)c_m \,\text{Im}\sigma, -M\right]\} \tag{2.1.17}$$

where $\delta > 0$ is however small and $M > 0$ is sufficiently large. If we demonstrate that, on the set (2.1.17), moduli of all factors of the product

$$V(\lambda, \sigma) = \left[V_0(0, 1) + \widetilde{\phi}(\lambda) V_1(0, 1)\right] \prod_{j=1}^{m} \left[\sigma - \sigma_j(\lambda)\right]$$

are greater than a positive constant, then we shall get the estimate (2.1.16).

Since $V_0(0, 1) \neq 0$, by virtue of boundedness of the normal surface for the operator V_0,

$$\left|V_0(0, 1) + \widetilde{\phi}(\lambda) V_1(0, 1)\right| \geq \text{const} > 0, \text{for Im}\lambda \leq -M$$

Furthermore, it follows from Lemma 2.1.1(b) that the inequalities

$$c_j \text{Im}\sigma_j(\lambda) \leq \text{Im}\lambda; \quad \text{for Im}\lambda \leq -M, \ j = 1, \ldots, m \tag{2.1.18}$$

hold. We have taken into account non-negativeness of the coefficients $k_j, j = 1, \ldots, m$ and the inequality $\text{Im} \left[\lambda \widetilde{\phi}(\lambda)\right] \leq 0$, which follows from Lemma 1.5.4. From the inequalities (2.1.18), we obtain that, on the set (2.1.17)

$$|\sigma - \sigma_j(\lambda)| \geq \frac{1}{|c_j|}\left|c_j\text{Im}\sigma - c_j\text{Im}\sigma_j(\lambda)\right|$$

$$= \frac{1}{|c_j|}\left|c_j\text{Im}\sigma - \text{Im}\lambda + \text{Im}\lambda - c_j\text{Im}\sigma_j(\lambda)\right|$$

$$\geq \frac{1}{|c_j|}\left|c_j\text{Im}\sigma - \text{Im}\lambda\right| \geq \text{const} > 0; \quad \in j = 1, \ldots, m$$

Thus the desired estimate (2.1.16) is obtained. A more precise description of the support of the fundamental solution $E(t, x)$ can be given with the help of Theorem 1.3.1. To make use of Theorem 1.3.1, we have to carry out some preliminary calculations.

Let us represent the function $[i^m V(\lambda, \sigma)]^{-1}$ as a sum of elementary fractions

$$\frac{1}{i^m V(\lambda, \sigma)} = \sum_{j=1}^{m} \frac{H_j(\lambda)}{\sigma - \sigma_j(\lambda)} \tag{2.1.19}$$

where

$$H_j(\lambda) \equiv \left\{ i^m \left[V_0(0, 1) + \tilde{\phi}(\lambda)V_1(0, 1)\right] \prod_{s, s \neq j} \left[\sigma_j(\lambda) - \sigma_s(\lambda)\right] \right\}^{-1}$$

$$= \left[i^m \frac{\partial}{\partial\sigma} V(\lambda, \sigma_j(\lambda)) \right]^{-1} \tag{2.1.20}$$

Note that by virtue of (2.1.18), the inequalities

$$\text{Im}\sigma_j(\lambda) \geq \text{const} > 0, \quad \text{if } c_j < 0 \tag{2.1.21}$$

$$\text{Im}\sigma_j(\lambda) \leq -\text{const} < 0, \quad \text{if } c_j > 0$$

hold for $- \text{Im} \lambda$ large enough. Hence

$$\tilde{E}(\lambda, \sigma) = F_{\sigma \to x}^{-1} F_{t, x \to \lambda, \sigma} E$$

$$= F_{\sigma \to x}^{-1} \frac{1}{i^m V(\lambda, \sigma)}$$

$$= \frac{1}{2\pi} \int_{-\infty}^{\infty} \sum_{j=1}^{m} \frac{H_j(\lambda)}{\sigma - \sigma_j(\lambda)} e^{ix\sigma} d\sigma \tag{2.1.22}$$

$$= i \sum_{j, c_j < 0} H_j(\lambda) e^{i\sigma_j(\lambda)x} \Theta(x) - i \sum_{j, c_j > 0} H_j(\lambda) e^{i\sigma_j(\lambda)x} \Theta(-x)$$

where

$$i = \sqrt{-1}, \quad \Theta(x) = \begin{cases} 1 & x > 0 \\ 0 & x < 0 \end{cases}$$

In what follows, we shall extend the definition of $\Theta(x)$ to $x = 0$ by putting

$$\Theta(0) = \frac{1}{2}$$

Clearly, in doing this, we do not change $E(t, x)$ regarded as a distribution in two variables t, x. From (2.1.22), it follows that $E(t, x)$ can be also considered a distribution in one variable t (while x is regarded as a parameter). It is evident that under such an approach $E(t, x)$ turns out to be infinitely differentiable with respect to the parameter x for $x \neq 0$. Moreover, $E(t, x)$ (regarded as a distribution in t) continuously depends on the parameter x, $-\infty < x < \infty$.

Really, as it follows from formula (2.1.22), it suffices to prove, for large $- \operatorname{Im} \lambda$, the equality

$$\sum_{j, c_j < 0} H_j(\lambda) = - \sum_{j, c_j > 0} H_j(\lambda)$$

That is

$$\sum_{j=1}^{m} H_j(\lambda) = 0$$

But

$$\sum_{j=1}^{m} H_j(\lambda) = \sum_{j=1}^{m} \left[i^m \frac{\partial}{\partial \lambda} V(\lambda, \sigma_j(\lambda)) \right]^{-1} = \frac{1}{2\pi i} \int_{\gamma} \frac{d\sigma}{i^m V(\lambda, \sigma)}$$

where γ is a closed counter-clockwisely oriented contour (in the complex σ-plane) containing inside of itself all the points $\sigma_j(\lambda)$. It is clear that the expression on the right-hand side of the last relation equals the coefficient to σ^{-1} in the Laurent's expansion of the function $[i^m V(\lambda, \sigma)]^{-1}$ about infinity. Therefore, the mentioned expression equals zero. Thus, the continuous dependence of $E(t, x)$ on the parameter x is established.

To be specific, suppose $x > 0$, $c_1 < 0$ and $c_1 < c_2 < \cdots < c_m$. Then for $- \operatorname{Im} \lambda$ large enough the above inequalities (2.1.18)

$$c_j \operatorname{Im} \sigma_j(\lambda) \leq \operatorname{Im} \lambda; j = 1, \ldots, m$$

yield the estimate

$$\left|\widetilde{E}(\lambda, x)\right| \leq \text{const} \exp\left(\frac{x}{|c_a|} \text{Im}\lambda\right), \quad \text{const} > 0$$

Hence by Theorem 1.3.1, the fundamental solution $E(t, x)$, regarded as a distribution in t, identically vanishes for $t < t(x) = x/|c_1|$, which leads us to the inclusion supp $E(t, x) \subseteq K$. Now, however, making use of the same Theorem 1.3.1, we can obtain new information about the support of $E(t, x)$. Namely, it will be demonstrated that the fundamental solution $E(t, x)$ regarded as a distribution in t does not vanish identically in however small neighbourhood of the point $t = t(x)$. Suppose the contrary, then, by Theorem 1.3.1, for some small $\delta > 0$, some $\nu > 0$ and sufficiently large $- \text{Im } \lambda$, we must have

$$\left|\widetilde{E}(\lambda, x)\right| \leq \text{const}(1 + |\lambda|)^{\nu} \exp\left(\frac{x}{|c_1| - \delta} \text{Im}\lambda\right), \text{const} > 0,$$

whence by virtue of (2.1.22), it follows that for p large enough, the inequality

$$\exp\left[-x\text{Im}\sigma_1(-ip)\right] \leq \text{const}(1 + p)^{\nu} \exp\left(-\frac{x}{|c_1| - \delta}p\right), \text{const} > 0,$$

must hold. Therefore, we must have

$$\text{Im}\sigma_1(-ip) \geq \frac{p}{|c_1| - \delta}[1 + o(1)], \quad p \to +\infty,$$

which evidently contradicts Lemma 2.1.1 (c).

Finally, it is clear that the point $P = (t(x), x)$ belongs to $\partial K \cap \partial \circ K$. The theorem is proved.

Notes About Supp $E(t, x)$ Supposing the hypotheses of the above theorem, let us give some supplements to the proof of sufficiency:

1. If the cone K coincides with the cone $\circ K$, then K evidently is the closure of the convex hull of supp $E(t, x)$.
2. Now suppose $K \supset \circ K$ (that is, K is a proper subset of $\circ K$); this means that all c_j, $j = 1, \ldots, m$ have the same sign. To be specific, we suppose $c_1 < \cdots < c_m < 0$.

Let us demonstrate that provided $V_1(0, 1) \neq 0$ and the function

$$\psi(t) = F_{\lambda \to t}^{-1} \frac{\widetilde{\phi}(\lambda)}{V_0(0, 1) + \widetilde{\phi}(\lambda) V_1(0, 1)}$$

has a non-compact support, the cone K is the closure of the convex hull of supp $E(t, x)$.

Suppose the contrary. Then there must exist $\delta > 0$, $\nu > 0$ such that

$$E(t,x) = 0, \quad \text{for } x \in [0,\delta], \quad t \in (-\infty, 0) \cup (\nu, +\infty),$$

whence it follows that

$$\lim_{x \to +0} \frac{\partial^{m-1} E(t,x)}{\partial x^{m-1}} = 0, \quad t \in (-\infty, 0) \cup (\nu, +\infty).$$

Furthermore, (2.1.22) gives

$$
\begin{aligned}
F_{t \to \lambda} \lim_{x \to +0} \frac{\partial^{m-1} E(t,x)}{\partial x^{m-1}} &= \lim_{x \to +0} \frac{\partial^{m-1} \widetilde{E}(\lambda,x)}{\partial x^{m-1}} \\
&= i \sum_{j=1}^{m} \frac{\left[i\sigma_j(\lambda)\right]^{m-1}}{i^m \frac{\partial}{\partial \sigma} V\left(\lambda, \sigma_j(\lambda)\right)} \qquad (2.1.23)\\
&= \frac{1}{2\pi i} \int_{\gamma} \frac{\sigma^{m-1}}{V(\lambda,\sigma)} \, d\sigma
\end{aligned}
$$

where γ is a closed counter-clockwisely oriented contour (in the complex σ-plane) containing inside of itself all the points $\sigma = \sigma_j(\lambda)$. The expression on the right-hand side of (2.1.23) evidently equals the coefficient to σ^{-1} in the Laurent's expansion of the function $\sigma^{m-1}/V(\lambda,\sigma)$ about infinity; as it is easy to check, the mentioned coefficient equals

$$\frac{1}{V_0(0,1) + \widetilde{\phi}(\lambda) V_1(0,1)} = \frac{V_1(0,1)}{V_0(0,1)} \left[\frac{1}{V_1(0,1)} - \widetilde{\psi}(\lambda) \right].$$

Therefore

$$\lim_{x \to +0} \frac{\partial^{m-1} E(t,x)}{\partial x^{m-1}} = \frac{\delta(t)}{V_0(0,1)} - \frac{\psi(t) V_1(0,1)}{V_0(0,1)}$$

whence it follows that the function $\psi(t)$ has a compact support, which gives the contradiction required.

3. Let $c_1 < \cdots < c_m < 0$ and suppose two following conditions

to be valid:

(a) The closure of the convex hull of supp $\psi(t)$ coincides with the segment $[0,\nu]$;
(b) $V_1(\lambda,\sigma) = \text{const } V_0(\lambda,\sigma)$.

Then supp $E(t,x) \subseteq Q$ (see Fig. 2.1). Moreover, by theorem of Lions [3], the closure of the convex hull of supp $E(t,x)$ coincides with Q.

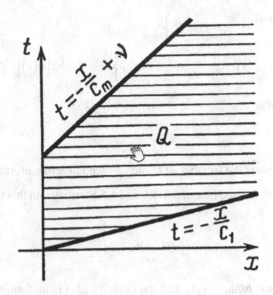

Fig. 2.1 Figure representing supp $E(t, x) \subseteq Q$

Note About Inhomogeneous Operators   Consider an operator with the symbol

$$V_0(\lambda, \sigma) + l_0(\lambda, \sigma) + \widetilde{\phi}(\lambda)[V_1(\lambda, \sigma) + l_1(\lambda, \sigma)] \tag{2.1.24}$$

where $V_i(\lambda, \sigma)$ are homogenous polynomials of degree m with real coefficient $l_i(\lambda, \sigma)$ are arbitrary polynomials of degree $\leq m - 1$; $i = 0, 1$. As before, we consider $V_0(\lambda, \sigma)$ a symbol of a strictly hyperbolic operator with a bounded normal surface (that is, in the decomposition (2.1.2), all c_j are different and distinct from zero). Let us write expansions of the polynomials $V_i(\lambda, \sigma)$ and $l_i(\lambda, \sigma)$ in powers of $(\lambda - c_1\sigma)$. Then the characteristic equation $V(\lambda, \sigma) = 0$ will assume the form

$$(\lambda - c_1\sigma)\left[a_{m-1}(\lambda - c_1\sigma)^{m-1} + \cdots + a_0\sigma^{m-1}\right]$$
$$+a_{m-1,1}(\lambda - c_1\sigma)^{m-1} + \cdots + a_{0,1}\sigma^{m-1} + \cdots$$
$$+\widetilde{\phi}(\lambda)\left[b_m(\lambda - c_1\sigma)^m + \cdots + b_0\sigma^m + b_{m,1}(\lambda - c_{1\sigma})^{m-1} + \cdots + b_{0,1}\sigma^{m-1} + \cdots\right] = 0$$

$$\tag{2.1.25}$$

The change (2.1.10) reduces the last equation to the form

$$z\left\{a_{m-1}\left[\widetilde{\phi}(\lambda)\right]^{m-1}z^{m-1}+\cdots+a_0\right\}$$

$$+\frac{1}{\lambda\widetilde{\phi}(\lambda)}\left\{a_{m-1,}\left[\widetilde{\phi}(\lambda)\right]^{m-1}z^{m-1}+\cdots+a_{0,1}\right\}+\cdots+b_m\left[\widetilde{\phi}(\lambda)\right]^{m}z^{m}+\cdots+b_0$$

$$+\frac{1}{\lambda}\left\{b_{m-1,1}\left[\widetilde{\phi}(\lambda)\right]^{m-1}z^{m-1}+\cdots+b_{0,1}+\cdots=0\right\}$$

$$(2.1.26)$$

Taking into account the fact that $\widetilde{\phi}(\lambda)$ and $\frac{1}{\lambda\widetilde{\phi}(\lambda)}$ tend to zero uniformly with respect to $\mathrm{Re}\lambda$ as $\mathrm{Im}\,\lambda \to -\infty$ (see Lemmas 1.5.1 and 1.5.8), we obtain that the Eq. (2.1.26) has a root

$$z_1 = -\frac{b_0}{a_0}+o(1)$$

where, as we know, $b_0/a_0 = c_1 k_1$ and the quantity $o(1)$ is uniform with respect to $\mathrm{Re}\lambda$. The corresponding root of Eq. (2.1.25) has the form

$$\sigma_1(\lambda) = \frac{\lambda}{c_1}+\frac{k_1+o(1)}{c_1}\lambda\widetilde{\phi}(\lambda)$$

Similarly, we obtain

$$\sigma_j(\lambda) = \frac{\lambda}{c_j}+\frac{k_j+o(1)}{c_j}\lambda\widetilde{\phi}(\lambda)$$

for the rest of j. All of these roots are continuous for $\mathrm{Im}\lambda \leq -M$ (provided $M > 0$ is sufficiently large). Now, in the manner of Theorem 2.1.1, we obtain the following result.

The operator (2.1.24) *is hyperbolic in S' and preserves this property under small real perturbances of the coefficients of operators V_i, $i = 0, 1$ and for small complex arbitrary perturbances of the coefficients of operators l_i, $i = 0, 1$ if and only if the inequalities*

$$k_j > 0; \quad j = 1, 2, \ldots, m$$

hold.

Further theorems of this chapter can also easily be generalized to the case of inhomogeneous operators. We leave it for the reader to do.

Now, we pass to the case where the function of memory has (as $t \to +0$) a singularity, which is not stronger than a logarithmic one.

Theorem 2.1.2 Let V_0 be a strictly hyperbolic operator with a bounded normal surface, and let

$$\phi(t) \leq \text{const} \ln \frac{1}{t}, \quad \text{const} > 0 \qquad (2.1.27)$$

for small $t > 0$. Then the operator V is hyperbolic in D', supp $E(t,x) \subseteq K$. And $E(t,x)$ does not identically wanish in a however small neighbourhood of an arbitrary point $P \in \partial K \cap \partial \circ K$. Here $E(t,x)$ is the operator V fundamental solution describing finite speed wave propagation; K and $\circ K$ are the influence cone and the propagation cone for the operator V_0, respectively.

Proof Consider a functional $E(t,x)$ defined on the space D of test functions (see Sect. 1.1.6) by the following integral:

$$\langle E(t,w), \psi(tx) \rangle = \frac{1}{(2\pi)^2} \int_{\bar{\Gamma}} d\lambda \int_{-\infty}^{\infty} \overline{\left[i^m V\left(\bar{\lambda}, \sigma\right) \right]^{-1}} F_{t,x \to \lambda, \sigma} \psi \, d\sigma \qquad (2.1.28)$$

or, which is the same

$$\langle E(t,x), \psi(tx) \rangle = \frac{1}{(2\pi)^2} \int_{\bar{\Gamma}} d\lambda \int_{-\infty}^{\infty} \overline{\sum_{j=1}^{m} \frac{H_j\left(\bar{\lambda}\right)}{\sigma - \sigma_j\left(\bar{\lambda}\right)}} F_{t,x \to \lambda, \sigma} \psi \, d\sigma \qquad (2.1.29)$$

Here the bar denotes the complex conjugation, functions H_j are defined by (2.1.20), and the contour $\bar{\Gamma}$ is defined by the equation

$$\text{Im} \lambda = M [\ln (|\text{Re} \lambda| + 1) + 1], \quad M > 0 \qquad (2.1.30)$$

Here we consider $M > 0$ sufficiently large. The contour orientation is given by the condition of coincidence of its projection on the Imλ- axis with the orientation of Imλ- axis.

First of all, let us prove the continuity of the functional $E(t,x)$. To do this, we have to study the behaviour of the quantities $\text{Im}\sigma_j(\lambda)$, $j = 1, \ldots, m$, when λ belongs to the contour Γ, which is defined by the equation

$$-\text{Im} \lambda = M [\ln (|\text{Re} \lambda| + 1) + 1] \qquad (2.1.31)$$

As usual, we denote $\sigma_j(\lambda)$ by the roots of the characteristic equation $V(\lambda, \sigma) = 0$. Consider, for example, the root $\sigma_1(\lambda)$. As we know from Lemma 2.1.1(b)

$$c_1 \operatorname{Im}\sigma_1(\lambda) = \operatorname{Im}\lambda + k_1[1 + o(1)] \operatorname{Im}\left[\lambda\widetilde{\phi}(\lambda)\right]$$

where the quantity $o(1)$ goes to zero uniformly with respect to $\operatorname{Re}\lambda$ as $\operatorname{Im}\lambda \to -\infty$.
Let us demonstrate that

$$c_1 \operatorname{Im}\sigma(\lambda)|_\Gamma \leq \text{const} < 0 \tag{2.1.32}$$

By Lemma 1.5.9, we have for $\operatorname{Im}\lambda \leq -\varepsilon < 0$:

$$0 \leq -\operatorname{Im}\left(\lambda\widetilde{\phi}(\lambda)\right) \leq a[\ln\left(|\operatorname{Re}\lambda| + 1\right) + \ln\left(|\operatorname{Im}\lambda| + 1\right) + 1]; \quad a > 0; \tag{2.1.33}$$

From Eq. (2.1.33), it follows that to prove the inequality (2.1.32), it suffices to demonstrate, for $\lambda \in \Gamma$, the inequality

$$|\operatorname{Im}\lambda| \geq |2k_1 a[\ln\left(|\operatorname{Re}\lambda| + 1\right) + \ln\left(|\operatorname{Im}\lambda| + 1\right) + 1]|.$$

But since by virtue of (2.1.30)

$$\ln\left(|\operatorname{Re}\lambda| + 1\right) = \frac{\operatorname{Im}\lambda}{M} - 1,$$

it suffices to demonstrate, for $\lambda \in \Gamma$, that the inequality

$$|\operatorname{Im}\lambda| \geq \left|2k_1 a\left[\frac{|\operatorname{Im}\lambda|}{M} + \ln\left(|\operatorname{Im}\lambda| + 1\right)\right]\right| \tag{2.1.34}$$

holds. However, it is clear that for $M > 0$ large enough, (2.1.34) will hold true (for arbitrary sign of k_1) in the half-plane. $\operatorname{Im}\lambda \leq -M$. But the contour Γ is located just in this half-plane. Thus the inequality (2.1.32) is established.

Similar constructions can be carried out for the rest of the roots $\sigma_j(\lambda)$. Therefore, one can choose $M > 0$ so large that the inequalities

$$c_j \operatorname{Im}\sigma_j(\lambda)|_\Gamma \leq \text{const} < 0; j = 1, \ldots, m \tag{2.1.35}$$

will hold true. But these inequalities yield the functional continuity on the space D of infinitely differentiable test functions with compact support. That is $E(t, x) \in D'$.

Now let us demonstrate that $E(t, x)$ is a fundamental solution for the operator V. In fact

$$\left\langle VE(t,x), \varphi(t,x) \right\rangle = \left\langle E(t,x), \left[(-1)^m V_0 \left(\frac{\partial}{\partial t}, \frac{\partial}{\partial x} \right) + \phi(-t) * (-1)^m V_1 \left(\frac{\partial}{\partial t}, \frac{\partial}{\partial x} \right) \right] \right.$$

$$\left. \varphi(t,x) \right\rangle = \frac{1}{(2\pi)^2} \int_{\overline{\Gamma}} d\lambda \int_{-\infty}^{\infty} \overline{\left(i^m V(\overline{\lambda}, \sigma) \right)^{-1}} \times \overline{i^m V(\overline{\lambda}, \sigma)} \, F_{t,x \to \lambda, \sigma} \, \varphi \, d\sigma$$

$$= \frac{1}{(2\pi)^2} \int_{\overline{\Gamma}} d\lambda \int_{-\infty}^{\infty} F_{t,x \to \lambda, \sigma} \varphi \, d\sigma = \frac{1}{(2\pi)^2} \int_{-\infty}^{\infty} d\lambda \int_{-\infty}^{\infty} F_{t,x \to \lambda, \sigma} \varphi \, d\sigma$$

$$= \varphi(0,0)$$

Here we use the well-known properties of convolution and the Cauchy theorem. But the last equality means that

$$VE(t,x) = \delta(t) \, \delta(x).$$

Now, let us study the support of the fundamental solution $E(t,x)$. To do this, we employ the Parseval's identity in the inner integral in (2.1.29)

$$\left\langle E(t,x), \varphi(t,x) \right\rangle = \frac{1}{2\pi} \int_{\overline{\Gamma}} d\lambda \int_{-\infty}^{\infty} \overline{F_{\sigma \to x}^{-1} \sum_{j=1}^{m} \frac{H_j(\overline{\lambda})}{\sigma - \sigma_j(\overline{\lambda})}} F_{\sigma \to x}^{-1} F_{t,x \to \lambda, \sigma}^{-1} \varphi \, dx \quad (2.1.36)$$

By inequalities (2.1.35), the following formula

$$F_{\sigma \to x}^{-1} \sum_{j=1}^{m} \frac{H_j(\lambda)}{\sigma - \sigma_j(\lambda)} = i \sum_{j, c_j < 0} H_j(\lambda) \, e^{i \sigma_j(\lambda) x} \Theta(x)$$

$$- i \sum_{j, c_j > 0} H_j(\lambda) \, e^{i \sigma_j(\lambda) x} \Theta(x) \quad (2.1.37)$$

$$\equiv \widetilde{E}_x(\lambda)$$

holds true for $\lambda \in \Gamma$ (that is, $\overline{\lambda} \in \overline{\Gamma}$) [see (2.1.22)]. Clearly, the function $\widetilde{E}_x(\lambda)$ can be analytically extended to some half-plane $-\,\mathrm{Im}\,\lambda \geq \mathrm{const} > 0$. To be specific, suppose $x > 0$, $c_1 < 0$ and $c_1 < \cdots < c_m$. Since for large $-\,\mathrm{Im}\,\lambda$, (2.1.7 and 2.1.33) yield the estimates

$$\left| c_j \mathrm{Im} \sigma_j(\lambda) - \mathrm{Im}\lambda \right| \leq \upsilon [\ln(|\lambda| + 1) + 1];$$

for $\upsilon = \mathrm{const} > 0$; $j = 1, \ldots, m$, we obtain that for large $-\,\mathrm{Im}\,\lambda$, the following inequality

$$\left|\widetilde{E}_x(\lambda)\right| \leq \text{const}\,(1+|\lambda|)^v \exp\left(\frac{x}{|c_1|}\text{Im}\lambda\right); \text{const} > 0$$

holds true, while none inequality of the type

$$\left|\widetilde{E}_x(\lambda)\right| \leq \text{const}\,(1+|\lambda|)^{v_1} \exp\left(\frac{x}{|c_1|-\delta}\text{Im}\lambda\right); \quad \delta > 0, \quad \text{const} > 0$$

is valid. Hence by Theorem 1.3.1, it follows that the function

$$E_x(t) = F_{\lambda \to t}^{-1}\widetilde{E}_x(\lambda)$$

vanishes for $t < t(x) = x/|c_1|$ and is not identically equal to zero in, however, small neighbourhood of the point $t = t(x)$.

Finally, interchanging integrals in (2.1.36) and using the Cauchy's theorem and Parseval's identity, we obtain for $M > 0$ large enough

$$\langle E(t,x), \varphi(t,x)\rangle = \frac{1}{2\pi}\int\limits_{-\infty}^{\infty}dx\int\limits_{\overline{\Gamma}}\overline{\widetilde{E}_x(\overline{\lambda})}\,F_{t\to\lambda}\varphi(t,x)d\lambda = \frac{1}{2\pi}\int\limits_{-\infty}^{\infty}dx$$

$$\int\limits_{iM-\infty}^{iM+\infty}\overline{\widetilde{E}_x(\overline{\lambda})}\,F_{t\to\lambda}\varphi(t,x)\,d\lambda = \int\limits_{-\infty}^{+\infty}\langle e^{-Mt}E_x(t), e^{Mt}\varphi(t,x)\rangle\,dx = \int\limits_{-\infty}^{+\infty}\langle E_x(t), \varphi(t,x)\rangle\,dx$$

Whence it follows that the fundamental solution $E(t, x)$ possesses all the required properties. The theorem is proved.

The following lemma presents a modification of a result from [4].

Lemma 2.1.2 Suppose for each $M \geq \text{const} > 0$, there exists a point $(\lambda_0(M), \sigma_0(M)) \in \mathbb{C}^1 \times \mathbb{C}^1$ whose coordinates satisfy the following conditions:

$$-\text{Im}\lambda_0(M) = [\ln\,(|\,\text{Re}\,\lambda_0(M)|+1)+1], \tag{2.1.38}$$

$$\text{Im}\,\sigma_0(M) = 0, \tag{2.1.39}$$

$$|\sigma_0(M)| \leq \gamma_1|\lambda_0(M)|^{\gamma_2}; \quad \gamma_1 = \text{const} > 0, \quad \gamma_2 = \text{const} > 0, \tag{2.1.40}$$

$$V(\lambda_0(M), \sigma_0(M)) = 0 \tag{2.1.41}$$

Then the operator V cannot be hyperbolic in D'.

Proof: Suppose the Contrary Let $E(t,x) \in D'$ be the operator V fundamental solution such that $\operatorname{supp} E(t,x) \subseteq Q$, where Q is a proper cone of the half-plane $t \geq 0$..

Following the method of [4], we consider an infinitely differentiable function $\psi(t, x)$, which has a compact support and equals unity in a neighbourhood of the origin of coordinates. Then

$$V[\psi(t,x)\,E(t,x)] = \delta(t)\,\delta(x) + g(t,x) \qquad (2.1.42)$$

where

$$g(t,x) \equiv V[(\psi(t,x) - 1)E(t,x)].$$

It is easy to see that distributions $\psi(t,x)E(t,x)$ and $g(t,x)$ have compact supports. Let

$$h(\eta_1, \eta_2) = \sup_{t,\,x \in \operatorname{supp} g(t,x)} (t\eta_1 + x\eta_2),\ \left(\eta_1 \in \mathbb{R}^1, \eta_2 \in \mathbb{R}^1\right).$$

It is clear that

$$h(s\eta_1, s\eta_2) = \operatorname{sh}(\eta_1, \eta_2), \quad s > 0,$$

and

$$h(-1, 0) = -A_1 < 0,$$

since $\operatorname{supp} g(t, x)$ is a compact subset of $Q \backslash \{0\}$. Now, let us apply the Fourier–Laplace transform to the equality (2.1.42). We have

$$(-i)^m V(\lambda, \sigma)H(\lambda, \sigma) = 1 + G(\lambda, \sigma) \qquad (2.1.43)$$

where

$$H(\lambda, \sigma) = F_{t,x \to \lambda, \sigma}(\psi E) \quad \text{and} \quad G(\lambda, \sigma) = F_{t,x \to \lambda, \sigma}g$$

are entire functions of λ, σ, as $\psi(t,x)\,E(t,x)$ and $g(t,x)$ have compact supports). Moreover, it follows from Theorem 1.3.3 that

$$|G(\lambda, \sigma)| \leq C(1 + |\lambda| + |\sigma|)^v\, e^{h(\operatorname{Im}\lambda,\operatorname{Im}\sigma)} \qquad (2.1.44)$$

where $C = \text{const} > 0$, $v = \text{const} > 0$. Furthermore, (2.1.41, 2.1.43, and 2.1.44) yield

$$h(\operatorname{Im}\lambda_0(M), \operatorname{Im}\sigma_0(M)) \geq -A_2[\ln(|\lambda_0(M)| + |\sigma_0(M)| + 1) + 1] \qquad (2.1.45)$$

where $A_2 = \text{const} > 0$. But from the condition (2.1.39), it follows that

$$\begin{aligned} h(\operatorname{Im}\lambda_0(M), \operatorname{Im}\sigma_0(M)) &= h(\operatorname{Im}\lambda_0(M), 0) \\ &= h(-|\operatorname{Im}\lambda_0(M)|, 0) \\ &= -A_1|\operatorname{Im}\lambda_0(M)| \end{aligned}$$

Hence (2.1.45) assumes the form

$$|\operatorname{Im}\lambda_0(M)| \leq \frac{A_2}{A_1}[\ln(|\lambda_0(M)| + |\sigma_0(M)| + 1) + 1]$$

whence on account of the condition (2.1.40), it follows that we must have the inequality

$$|\operatorname{Im}\lambda_0(M)| \leq A[\ln(|\operatorname{Im}\lambda_0(M)| + |\operatorname{Re}\lambda_0(M)| + 1) + 1] \qquad (2.1.46)$$

where $A > 0$ is some constant, which does not depend on M. Finally, (2.1.46) yields the inequality

$$|\operatorname{Re}\lambda_0(M)| \geq \exp\left(\frac{|\operatorname{Im}\lambda_0(M)|}{A} - 1\right) - |\operatorname{Im}\lambda_0(M)| - 1$$

which evidently contradicts the condition (2.1.38) for $M > 0$ sufficiently large. The lemma is proved.

Note In proving Lemma 2.1.2, we did not employ the special form of the operator V (in fact, we employed only the symbol $V(\lambda, \sigma)$ being holomorphic for $\operatorname{Im}\lambda \leq \text{const}$). Therefore, Lemma 2.1.2 will hold true if we replace the operator V by W with the symbol (1.1.34).

Theorem 2.1.3 Let V_0 be a strictly hyperbolic operator with a bounded normal surface and let

$$\overline{\lim_{t\to+0}} \frac{\phi(t)}{\ln(1/t)} = +\infty \qquad (2.1.47)$$

Then all the assertions of Theorem 2.1.1 hold true for the space D'.

Proof In essence, we have to prove only the fact that the operator V cannot be hyperbolic in D' if among $k_j, j = 1, \ldots, m$, there exists even though one negative.

Let, for example, $k_1 < 0$. Consider the root $\sigma_1(\lambda)$ of the characteristic equation $V(\lambda, \sigma) = 0$. As we know from Lemma 2.1.1(b)

$$c_1 \operatorname{Im} \sigma_1(\lambda) = \operatorname{Im} \lambda + k_1 \left[1 + o(1)\right] \operatorname{Im} \left[\lambda \widetilde{\phi}(\lambda)\right]$$

where the quantity $o(1)$ goes to zero uniformly with respect to $\operatorname{Re} \lambda$, as $\operatorname{Im} \lambda \to -\infty$. Let us study the sign of $c_1 \operatorname{Im} \sigma_1(\lambda)$ on the contour Γ given by the equation

$$-\operatorname{Im} \lambda = M \left[\ln \left(|\operatorname{Re} \lambda| + 1\right) + 1\right]$$

where $M > 0$ is sufficiently large. At first let $\operatorname{Re} \lambda = 0$, $\lambda \in \Gamma$, then $\lambda = -iM$ and, as we know

$$c_1 \operatorname{Im} \sigma_1 (-iM) < 0 \tag{2.1.48}$$

for $M > 0$ large enough. Now let $\operatorname{Re} \lambda \to +\infty$, then we evidently have

$$\frac{\operatorname{Re} \lambda}{\operatorname{Im} \lambda} \le -\text{const} < 0$$

on the contour Γ. Therefore, on account of Lemma 1.5.4, for $\operatorname{Re} \lambda \to +\infty$, $\lambda \in \Gamma$, the following inequality holds true:

$$-\operatorname{Im} \left[\lambda \widetilde{\phi}(\lambda)\right] \ge \text{const} \, \phi \left(\frac{\pi}{2 \operatorname{Re} \lambda}\right), \quad \text{const} > 0$$

Hence taking into account the condition (2.1.47) and the fact that $k_1 < 0$, we obtain that, for $\operatorname{Re} \lambda$ large enough, on the contour Γ, there do exist points λ such that

$$c_1 \operatorname{Im} \sigma_1(\lambda) < 0 \tag{2.1.49}$$

Thus by virtue of continuity of the root $\sigma_1(\lambda)$, it follows from (2.1.48, 2.1.49) that on the contour Γ, there exists a point $\lambda_0 = \lambda_0(M)$ for which

$$\operatorname{Im} \sigma_1 \left[\lambda_0(M)\right] = 0. \tag{2.1.50}$$

Besides that, it is evident from Lemma 2.1.1(a) that for $M > 0$ large enough

$$|\sigma_1(\lambda_0(M))| \le \frac{2|\lambda_0(M)|}{c_1}. \tag{2.1.51}$$

Letting $\sigma_0(M) = \sigma_1(\lambda_0(M))$, we obtain from (2.1.50, 2.1.51) that the coordinates of the point $(\lambda_0(M), \sigma_0(M))$ satisfy all the conditions of Lemma 2.1.2 for $M > 0$ large enough, which gives the result required.

2.2 Multiple Characteristics, Multiple Convolutions, Unbounded Normal Surface

This section is devoted to some generalizations of Theorems 2.1.1, 2.1.2, and 2.1.3.

Theorem 2.2.1 Let W be an operator with the symbol

$$W(\lambda, \sigma) = \sum_{s=0}^{\infty} \left[\widetilde{\phi}(\lambda) \right]^s V_s(\lambda, \sigma); \quad \lambda \in C^1, \ \sigma \in C^1. \tag{2.2.1}$$

For the operator W to be hyperbolic in D', it is necessary that two following conditions hold:

(a) If $\lambda - c_j\sigma, c_j \neq 0$ is a multiplier of multiplicity q_j of $V_0(\lambda, \sigma)$, then $\lambda - c_j\sigma$ is also a multiplier of multiplicity $\geq \max{(q_j - s, 0)}$ of $V_s(\lambda, \sigma)$; $s = 1, 2, \ldots$,
(b) If λ is a multiplier of multiplicity q of $V_0(\lambda, \sigma)$, then λ is also a multiplier of multiplicity $\geq q$ of $V_s(\lambda, \sigma)$; $s = 1, 2, \ldots$

Proof For simplicity, we suppose $V_s(\lambda, \sigma) = 0$ for $s \geq 3$. (The general case can be treated similarly.)

(a) Let $\lambda - c_j\sigma, c \neq 0$, be a multiplier of multiplicity q_j of the polynomial $V_0(\lambda, \sigma)$ (for ease of notation, we shall consider $j = 1$). Then polynomials $V_s(\lambda, \sigma)$ can be represented (in a unique way) in the following form:

$$V_0(\lambda, \sigma) = (\lambda - c_1\sigma)^{q_1} \left[a_{m-q_1}(\lambda - c_1\sigma)^{m-q_1} + \right.$$
$$\left. + \cdots + a_0\sigma^{m-q_1} \right], \quad a_0 \neq 0,$$
$$V_1(\lambda, \sigma) = b_m(\lambda - c_1\sigma)^m + b_{m-1}(\lambda - c_1\sigma)^{m-1}\sigma + \cdots + b_0\sigma^m,$$
$$V_2(\lambda, \sigma) = d_m(\lambda - c_1\sigma)^m + d_{m-1}(\lambda - c_1\sigma)^{m-1}\sigma + \cdots + d_0\sigma^m,$$

so the characteristic equation assumes the form

$$W(\lambda, \sigma) \equiv (\lambda - c_1\sigma)^{q_1} \left[a_{m-q_1}(\lambda - c_1\sigma)^{m-q_1} + \cdots + a_0\sigma^{m-q_1} \right]$$
$$+ \widetilde{\phi}(\lambda)[b_m(\lambda - c_1\sigma)^m + \cdots + b_0\sigma^m] \tag{2.2.2}$$
$$+ \left[\widetilde{\phi}(\lambda) \right]^2 [d_m(\lambda - c_1\sigma)^m + \ldots + d_0\sigma^m] = 0$$

First of all, let us introduce a change

$$y = \frac{\lambda - c_1\sigma}{\sigma}, \tag{2.2.3}$$

Then (2.2.2) assumes the form

$$y_1^{q_1}\left(a_{m-q_1}y^{m-q_1}+\cdots+a_0\right)+\widetilde{\phi}(\lambda)(b_my^m+\cdots+b_0)$$
$$+\left[\widetilde{\phi}(\lambda)\right]^2(d_my^m+\cdots+d_0)=0. \tag{2.2.4}$$

Let b_x and d_k be the lowest-order nonzero coefficients in the corresponding brackets in (2.2.4) (that is, $0=b_{x-1}=b_{x-2}=\cdots,\ 0=d_{k-1}=d_{k-2}=\cdots$). Let us show that a necessary condition for the operator W to be hyperbolic is

$$x\geq q_1-1;\quad k\geq q_1-2$$

Suppose the contrary. Namely, let even if one of two inequalities

$$x<q_1-1;k<q_1-2 \tag{2.2.5}$$

be a valid.

To be specific, we consider both inequalities (2.2.5) to be performed. Let us define the number w by the condition

$$q_1w=\min\{1+xw,2+kw\}.$$

It is clear that under the conditions (2.2.5), the solution of the last equation is given by formula

$$w=\min\left\{\frac{1}{q_1-x},\frac{2}{q_1-k}\right\}$$

whence it follows that w is a rational number which can be represented as an irreducible fraction N_1/N_2 and $0<w<1$.

Let us introduce one more change:

$$z=\frac{y}{\left[\widetilde{\phi}(\lambda)\right]^w}, \tag{2.2.6}$$

then Eq. (2.2.4) will take the form

$$\left[\widetilde{\phi}(\lambda)\right]^{q_1w}z^{q_1}\left[\cdots+a_0\right]+\widetilde{\phi}(\lambda)\left\{\cdots+b_x\left[\widetilde{\phi}(\lambda)\right]^{xw}z^x\right\}$$
$$+\left[\widetilde{\phi}(\lambda)\right]^2\left\{\cdots+d_k\left[\widetilde{\phi}(\lambda)\right]^{kw}z^k\right\}=0. \tag{2.2.7}$$

By cancelling $\left[\widetilde{\phi}(\lambda)\right]^{q_1w}$, we have, as $\widetilde{\phi}(\lambda)\to0$ (that is, as $\mathrm{Im}\lambda\to-\infty$), the limiting equation for $z=z(\lambda)$ has one of three following forms:

(a) $a_0 z^{q_1} + b_x z^x = 0$;

(b) $a_0 z^{q_1} + d_k z^k = 0$;

(c) $a_0 z^{q_1} + b_x z^x + d_k z^k = 0, \; x > k$

where $a_0, b_x, d_k \neq 0..$

Suppose, for example, the case (c) takes place. To be specific, consider the case where the limiting equation has two complex conjugated roots Z_1 and $Z_2 = \overline{Z}_1$.

Therefore, for $- \operatorname{Im} \lambda$ large enough, Eq. (2.2.7) has continuous solutions of the form

$$z_1(\lambda) = Z_1 + o(1), z_2(\lambda) = Z_2 + o(1)$$

whence it follows that the original Eq. (2.2.2) has continuous solutions of the form

$$\sigma_1 = \frac{\lambda}{c_1 + z_1(\lambda)\left[\widetilde{\phi}(\lambda)\right]^w} = \frac{\lambda}{c_1} - \frac{Z_1 + o(1)}{c_1} \lambda \left[\widetilde{\phi}(\lambda)\right]^w \qquad (2.2.8)$$

$$\sigma_2 = \frac{\lambda}{c_1 + z_2(\lambda)\left[\widetilde{\phi}(\lambda)\right]^w} = \frac{\lambda}{c_1} - \frac{Z_2 + o(1)}{c_1} \lambda \left[\widetilde{\phi}(\lambda)\right]^w \qquad (2.2.9)$$

Here the quantity $o(1)$ goes to zero uniformly with respect to $\operatorname{Re}\lambda$, as $\operatorname{Im}\lambda \to - \infty$).

Let us begin with the case where in the representation of the number w as an irreducible fraction, N_1/N_2 the denominator $N_2 \geq 3$.

Let us study the sign of $c_1 \operatorname{Im} \sigma_1(\lambda)$ on the contour Γ given by the equation

$$-\operatorname{Im}\lambda = M[\ln\left(|\operatorname{Re}\lambda| + 1\right) + 1], \; M > 0. \qquad (2.2.10)$$

At first, suppose $\operatorname{Re}\lambda = 0 \, (\lambda \in \Gamma)$, then $\lambda = - iM$, and as we know,

$$c_1 \operatorname{Im}\sigma_1(-iM) < 0 \qquad (2.2.11)$$

provided M is sufficiently large.

Furthermore, from Lemma 1.5.7, it follows that for $\operatorname{Re}\lambda \to - \infty$

$$\arg\widetilde{\phi}(\lambda)\Big|_\Gamma \in \left[\theta_0, \frac{\pi}{2}\right] \qquad (2.2.12)$$

where $0 < \theta_0 < \pi/2$. It is geometrically evident that one can always choose the function $\left[\widetilde{\phi}(\lambda)\right]^w$ branch such that two following conditions

$$\left| \frac{\operatorname{Im}\left\{ Z_1 \lambda \left[\widetilde{\phi}(\lambda) \right]^w \right\}}{\operatorname{Re}\left\{ Z_1 \lambda \left[\widetilde{\phi}(\lambda) \right]^w \right\}} \right| \geq \text{const} > 0; \quad \lambda \in \Gamma, \quad \operatorname{Re}\lambda \to -\infty, \qquad (2.2.13)$$

$$\operatorname{Im}\left\{ -Z_1 \lambda \left[\widetilde{\phi}(\lambda) \right]^w \right\} > 0; \quad \lambda \in \Gamma, \quad \operatorname{Re}\lambda \to -\infty \qquad (2.2.14)$$

hold. The condition (2.2.13) yields

$$\operatorname{Im}\lambda = o\left(\operatorname{Im}\left\{ [1 + o(1)] Z_1 \lambda \left[\widetilde{\phi}(\lambda) \right]^w \right\} \right); \lambda \in \Gamma, \operatorname{Re} \to -\infty \qquad (2.2.15)$$

Really, on the one hand

$$|\operatorname{Im}\lambda| \leq \text{const} \,|\ln\,(|\lambda| + 1) + 1|; \quad \lambda \in \Gamma, \quad \operatorname{Re}\lambda \to -\infty$$

by virtue of Eq. (2.2.10). On the other hand, by Eq. (2.2.13), for $\lambda \in \Gamma$, $\operatorname{Re}\lambda \to -\infty$,

$$\left| \operatorname{Im}\left\{ [1 + o(1)] Z_1 \lambda \left[\widetilde{\phi}(\lambda) \right]^w \right\} \right| \geq \text{const} \left| \lambda^{1-w} \left[\lambda \widetilde{\phi}(\lambda) \right]^w \right|$$

$$\geq \text{const} \,|\lambda|^{1-w}, \quad \text{const} > 0,$$

since by Lemma 1.5.8

$$\left| \lambda \widetilde{\phi}(\lambda) \right| \to \infty, \quad \text{for} \lambda \in \Gamma, \quad \operatorname{Re}\lambda \to -\infty.$$

Thus the relation (2.2.15) is proved. Now, from (2.2.8, 2.2.14, and 2.2.15), it is clear that

$$c_1 \operatorname{Im}\sigma_1(\lambda)|_\Gamma > 0, \quad \text{as } \operatorname{Re}\lambda \to -\infty \qquad (2.2.16)$$

Thus from (2.2.11, 2.2.16) by virtue of continuity of $\sigma_1(\lambda)$, it follows that there exists a point $\lambda_0 = \lambda_0(M) \in \Gamma$ such that

$$\operatorname{Im}\sigma_1[\lambda_0(M)] = 0 \qquad (2.2.17)$$

Besides that, from (2.2.8), it is evident that

$$|\sigma_1[\lambda_0(M)]| \leq \frac{2|\lambda_0(M)|}{|c_1|} \qquad (2.2.18)$$

for M large enough. Letting $\sigma_0(M) = \sigma_1[\lambda_0(M)]$, we obtain from (2.2.17, 2.2.18) that for sufficiently large M, the coordinates of the point $(\lambda_0(M), \sigma_0(M))$ satisfy all the conditions of Lemma 2.1.2 (where one should replace the operator V by W), whence it follows that the operator W cannot be hyperbolic in D'.

Consider now the case where in the representation $w = N_1/N_2$, the denominator $N_2 = 2$. Then $w = 1/2$ (since, as we know, $0 < w < 1$).

Without loss of generality, we consider

$$\arg Z_1 \in \left[-\pi, -\frac{\pi}{2}\right] \cup \left[0, \frac{\pi}{2}\right] \tag{2.2.19}$$

If this condition is not performed, one should replace Z_1 by $Z_2 = \overline{Z}_1$. From (2.2.12, 2.2.19) and the relation

$$\arg \lambda|_\Gamma \to -\pi, \text{ as } \operatorname{Re} \lambda \to -\infty$$

it is geometrically evident that one of two branches of the function $\left[\widetilde{\phi}(\lambda)\right]^{1/2}$ will satisfy the conditions (2.2.13, 2.2.14). Hence, as before, it follows that the operator W cannot be hyperbolic in D'.

(b) Let λ be a multiplier of multiplicity q of the polynomial $V_0(\lambda, \sigma)$. Then the characteristic equation $W(\lambda, \sigma) = 0$ can be written in the form analogous to (2.2.2)

$$\lambda^q \left(a_{m-q} \lambda^{m-q} + a_{m-q-1} \lambda^{m-q-1} \sigma + \cdots + a_0 \sigma^{m-q}\right)$$
$$+ \widetilde{\phi}(\lambda) \left(b_m \lambda^m + b_{m-1} \lambda^{m-1} \sigma + \cdots + b_0 \sigma^m\right)$$
$$+ \left[\widetilde{\phi}(\lambda)\right]^2 \left(d_m \lambda^m + d_{m-1} \lambda^{m-1} \sigma + \cdots + d_0 \sigma^m\right)$$
$$= 0. \tag{2.2.20}$$

Let b_x and d_k be the lowest-order nonzero coefficients in the corresponding brackets in (2.2.20). Let us show that for the operator W to be hyperbolic, it is necessary that $x \geq q$; $k \geq q$.

Suppose the Contrary Namely, let even if one of two inequalities

$$x < q; \quad k < q \tag{2.2.21}$$

be valid. To be specific, we consider both inequalities (2.2.21) to be performed. As before, let us define the number w by the condition

$$qw = \min \{1 + xw, \ 2 + kw\}.$$

It is clear that inequalities (2.2.21) if satisfied, then

$$w = \min\left\{\frac{1}{q-x}, \frac{2}{q-k}\right\} > 0$$

Now, choosing the principal branch of the function $\left[\tilde{\phi}(\lambda)\right]^w$ and making use of the changes (2.2.3, 2.2.6), we easily obtain that for $-\operatorname{Im}\lambda$ large enough, the Eq. (2.2.20) has a continuous solution of the form

$$\sigma(\lambda) = [\text{const} + o(1)]\frac{\lambda}{\left[\tilde{\phi}(\lambda)\right]^w}, \quad \text{const} \neq 0 \qquad (2.2.22)$$

where $o(1)$ goes to zero uniformly with respect to $\operatorname{Re}\lambda$, as $\operatorname{Im}\lambda \to -\infty$.

Note that by virtue of Lemma 1.5.8,

$$|\sigma(\lambda)| \leq \text{const}\left|\frac{\lambda}{\left[\tilde{\phi}(\lambda)\right]^w}\right| = \text{const}\frac{|\lambda|^{1+w}}{\left|\lambda\tilde{\phi}(\lambda)\right|^w} \leq \text{const}|\lambda|^{1+w} \qquad (2.2.23)$$

for $-\operatorname{Im}\lambda$ sufficiently large. Furthermore, from Lemma 1.5.7, it follows that

$$\arg\left[\tilde{\phi}(\lambda)\right]^w\Big|_\Gamma \in \left[-\frac{\pi w}{2}, -\theta_0 w\right], \quad \text{as } \operatorname{Re}\lambda \to +\infty,$$

or

$$\arg\left[\tilde{\phi}(\lambda)\right]^w\Big|_\Gamma \in \left[\theta_0 w, \frac{\pi w}{2}\right], \quad \text{as } \operatorname{Re}\lambda \to -\infty, \qquad (2.2.24)$$

where $0 < \theta_0 < \pi/2$. Taking into account the relations

$$\arg\lambda|_\Gamma \to 0, \quad \text{as } \operatorname{Re}\lambda \to +\infty,$$
$$\arg\lambda|_\Gamma \to -\pi, \quad \text{as } \operatorname{Re}\lambda \to -\infty,$$

one can easily obtain from (2.2.24) that

$$\arg\frac{\lambda}{\left[\tilde{\phi}(\lambda)\right]^w}$$

assumes, by continuity, each interim value from the interval $(-\pi - \theta w, \theta_0 w)$. Hence $\arg\sigma(\lambda)|_\Gamma$ also assumes each interim value from some interval of length $\pi + 2\theta_0 w > \pi$. Therefore, there exists a point $\lambda_0 = \lambda_0(M) \in \Gamma$ such that

$$\text{Im}\sigma\,[\lambda_0(M)] = 0 \qquad\qquad (2.2.25)$$

Letting $\sigma_0(M) = \sigma[\lambda_0(M)]$, we obtain from (2.2.23, 2.2.25) that for M large enough, the coordinates of the point $(\lambda_0(M), \sigma_0(M))$ satisfy all conditions of Lemma 2.1.2 (where one should replace the operator V by W), whence it follows that the operator W cannot be hyperbolic in D'. The theorem is proved.

The following theorem presents a natural generalization of a result known for differential hyperbolic operators (see [5, 6]).

Theorem 2.2.2 Let

$$\phi \le \text{const}\,\ln\frac{1}{t}$$

for small $t > 0$. Then the conditions of the previous theorem are also sufficient for the operator W to be hyperbolic (in D'). These conditions, if satisfied, supp $E(t, x) \subseteq K$. Here $E(t, x)$ is the operator W fundamental solution describing finite speed wave propagation; K is the influence cone for the operator V_0.

Proof The proof of this theorem is similar, in the main, to the one of Theorem 2.1.2. At first, let V_0 be an operator with a bounded normal surface. It is easy to show that in the case considered, the roots of the characteristic equation $W(\lambda, \sigma) = 0$ have the following form as $\text{Im}\lambda \to -\infty$

$$\sigma_j(\lambda) = \frac{\lambda}{c_j} + \frac{d_j + o(1)}{c_j}\lambda\widetilde{\phi}(\lambda); \quad j = 1, \ldots, m.$$

Here the quantity $o(1)$ is uniformly small with respect to $\text{Re}\lambda$.

For simplicity we suppose $c_1 < c_2 < \cdots < c_m < 0$. Then one can write the following formula, analogous to (2.1.37):

$$\widetilde{E}_x(\lambda) = \frac{\det\begin{vmatrix} 1 & \cdots & 1 \\ \sigma_1(\lambda) & \cdots & \sigma_m(\lambda) \\ \vdots & & \vdots \\ \sigma_1^{m-2}(\lambda) & \cdots & \sigma_m^{m-2}(\lambda) \\ e^{i\sigma_1(\lambda)x} & \cdots & e^{i\sigma_m(\lambda)x} \end{vmatrix}}{i^m\left\{\sum\limits_{s=0}^{\infty} V_s(0,1)\left[\widetilde{\phi}(\lambda)\right]^s\right\}\prod\limits_{l,\tau,l<\tau}\left[\sigma_l(\lambda) - \underset{\tau}{\sigma}(\lambda)\right]}\,\Theta(x); i = \sqrt{-1} \qquad (2.2.26)$$

Let us explain, how one should understand formula (2.2.26) in case under consideration when the operator V_0 has multiple characteristics and some of the roots $\sigma_j(\lambda)$, $j = 1, \ldots, m$ may coincide. Note that for $-\text{Im }\lambda$ large enough, from coincidence of the roots σ_j and σ_k, it follows that $c_j = c_k$. Suppose, for example, that for $\lambda = \lambda_0$, the roots $\sigma_1(\lambda)$, $\sigma_2(\lambda)$, $\sigma_3(\lambda)$ do coincide, and the rest of the roots are different and distinct from $\sigma_1(\lambda)$, $\sigma_2(\lambda)$, $\sigma_3(\lambda)$. First of all, let us demonstrate that the numerator of the fraction (2.2.26), that is, the determinant of the matrix

$$
\begin{vmatrix}
1 & \cdots & 1 \\
\sigma_1(\lambda) & \cdots & \sigma_m(\lambda) \\
\vdots & & \vdots \\
\sigma_1^{m-2}(\lambda) & \cdots & \sigma_m^{m-2}(\lambda) \\
e^{i\sigma_1(\lambda)x} & \cdots & e^{i\sigma_m(\lambda)x}
\end{vmatrix}
\tag{2.2.27}
$$

has, for $\lambda = \lambda_0$, a zero of order greater than or equal to the zero order of the denominator of the fraction considered. Let us subtract the first column of the matrix (2.2.27) from the second and third ones and then the second resulting column from the third one. Now one can single out of the determinant in the numerator of (2.2.26) the product

$$
(-1)^3 [\sigma_1(\lambda) - \sigma_2(\lambda)] [\sigma_1(\lambda) - \sigma_3(\lambda)] [\sigma_2(\lambda) - \sigma_3(\lambda)].
\tag{2.2.28}
$$

Besides that, the elements of the second and third columns of the resulting matrix (under determinant) will present divided differences[2] of first- and second-order, respectively. It is easy to show that these divided differences are bounded functions of λ in a neighbourhood of λ_0 [2], p. 451.

Now, let us cancel the numerator and the denominator in (2.2.26) by the product (2.2.28). We shall consider the function $\widetilde{E}_x(\lambda)$ equal to the result of such a cancellation. In case where several roots or several groups of roots do coincide, formula (2.2.26) should be interpreted similarly.

Thus the function $\widetilde{E}_x(\lambda)$ proves to be holomorphic for $-\operatorname{Im}\lambda$ sufficiently large. Using the representation of divided differences in the form of contour integrals [2], one can easily demonstrate that for $x > 0$ the function $\widetilde{E}_x(\lambda)$ satisfies the estimate

---

[2]The expression

$$
\Delta^{(1)}(f; \sigma_1, \sigma_2) = \frac{f(\sigma_2) - f(\sigma_1)}{\sigma_2 - \sigma_1}
$$

is called the *first-order divided difference* for the function $f(\sigma)$ with respect to points σ_1 and σ_2; the expression

$$
\Delta^{(2)}(f; \sigma_1, \sigma_2, \sigma_3) = \frac{\Delta^{(1)}(f; \sigma_2, \sigma_3) - \Delta^{(1)}(f; \sigma_1, \sigma_2)}{\sigma_3 - \sigma_1}
$$

is called the *second-order divided difference* for the function $f(\sigma)$ with respect to points σ_1, σ_2, σ_3, etc.

$$\left|\widetilde{E}_x(\lambda)\right| \le \text{const}\,(1+|\lambda|)^v\, e^{\frac{x}{|c_1|}\text{Im}\lambda}; \text{const} > 0, \quad v = \text{const} > 0$$

Now the assertion of the theorem follows from Theorem 1.3.1.

Finally, the case of an unbounded normal surface reduces to the case of a bounded normal surface by virtue of Theorem 2.2.1. The theorem is proved.

Let us pass to the case where

$$\varliminf_{t\to+0} \frac{\phi(t)}{\ln(1/t)} = +\infty$$

We shall demonstrate that in this case, even under necessary conditions of Theorem 2.2.1, the operator W hyperbolicity is not determined by operators V_0 and V_1, provided the characteristics of V_0 are multiple.

Let us consider, as an example, the operator whose symbol has the form

$$W(\lambda,\sigma) = (\lambda - c_1\sigma)^2 \left[a_{m-2}(\lambda - c_1\sigma)^{m-2} + a_{m-3}(\lambda - c_1\sigma)^{m-3}\sigma + \cdots + a_0\sigma^{m-2} \right]$$

$$+ \widetilde{\phi}(\lambda)(\lambda - c_1\sigma)\left[b_{m-1}(\lambda - c_1\sigma)^{m-1} + b_{m-2}(\lambda - c_1\sigma)^{m-2}\sigma + \ldots + b_0\sigma^{m-1} \right]$$

$$+ \left(\widetilde{\phi}(\lambda)\right)^2 \left[d_m(\lambda - c_1\sigma)^m + d_{m-1}(\lambda - c_1\sigma)^{m-1}\sigma + \cdots + d_0\sigma^m \right]$$

$$\text{(2.2.29)}$$

where $c_1 \ne 0$, $a_0 \ne 0$. Letting

$$Z = \frac{\lambda - c_1\sigma}{\sigma\widetilde{\phi}(\lambda)}$$

we easily obtain that the characteristic equation $W(\lambda,\sigma) = 0$ assumes the form

$$Z^2\left[a_{m-2}\left(\widetilde{\phi}(\lambda)\right)^{m-2} Z^{m-2} + \cdots + a_0 \right] + Z\left[b_{m-1}\left(\widetilde{\phi}(\lambda)\right)^{m-1} Z^{m-1} + \cdots + b_0 \right]$$

$$+ d_m\left(\widetilde{\phi}(\lambda)\right)^m Z^m + \cdots + d_0 = 0$$

For $-\text{Im}\,\lambda$ large enough, the last equation has, in particular, two roots

$$Z_{1,2}(\lambda) = \frac{-b_0 \pm \sqrt{b_0^2 - 4a_0d_0}}{2a_0} + o(1),$$

therefore, the characteristic equation $W(\lambda,\sigma) = 0$ has two continuous roots

$$\sigma_{1,2}(\lambda) = \frac{\lambda}{c_1} + \left[\frac{b_0 \pm \sqrt{b_0^2 - 4a_0d_0}}{2a_0c_1} + o(1)\right] \lambda\widetilde{\phi}(\lambda)$$

Here the quantity $o(1)$ tends to zero uniformly with respect to $\mathrm{Re}\lambda$, as $\mathrm{Im}\,\lambda \to -\infty$. Hence conditions of hyperbolicity for operators of the (2.2.29) type present concordance conditions for coefficients of three operators V_0, V_1 and V_2. The complexity of these conditions evidently grows with the multiplicity of roots and convolutions.

Now let us consider the case where the polynomial V_0 has a root of multiplicity 1. Then, as is easy to see, the following result generalizing Lemma 2.1.1 holds.

Lemma 2.2.1 Suppose in the decomposition

$$V_0(\lambda, \sigma) = V_0(1, 0) \prod_{j=1}^{m} (\lambda - c_j\sigma)$$

the coefficient c_{j_0} is distinct from zero and from all $c_j, j \neq j_0$. Then for $-\,\mathrm{Im}\,\lambda$ large enough, the characteristic equation $W(\lambda, \sigma) = 0$ has a continuous root $\sigma_{j_0}(\lambda)$ such that

(a) Uniformly with respect to $\mathrm{Re}\lambda$

$$\sigma_{j_0}(\lambda) = \frac{\lambda}{c_{j_0}} + \frac{k_{j_0} + o(1)}{c_{j_0}} \lambda\widetilde{\phi}(\lambda); k_{j_0} \equiv \left.\frac{V_1(\lambda, \sigma)}{\lambda\frac{\partial}{\partial\lambda}V_0(\lambda, \sigma)}\right|_{\lambda = c_{j_0}\sigma, \sigma \neq 0} ;$$

where $\mathrm{Im}\lambda \to -\infty$.

(b) Uniformly with respect to $\mathrm{Re}\lambda$

$$\mathrm{Im}\sigma_{j_0}(\lambda) = \frac{\mathrm{Im}\lambda}{c_{j_0}} + \frac{k_{j_0} + o(1)}{c_{j_0}} \mathrm{Im}\left[\lambda\widetilde{\phi}(\lambda)\right]; \mathrm{Im}\lambda \to -\infty$$

(c) For real p,

$$\mathrm{Im}\sigma_{j_0}(-ip) = -\frac{p}{c_{j_0}}(1 + o(1)); p \to +\infty$$

From Lemma 2.2.1, it evidently follows that for the operator W to be hyperbolic in S', it is necessary that

$$k_{j_0} \geq 0.$$

Now, if $V_0(\lambda, \sigma)$ is strictly hyperbolic, then Lemma 2.2.1 and Theorem 2.2.1 yield the following description of hyperbolic operators with memory.

Theorem 2.2.3 Let V_0 be a strictly hyperbolic operator.

1. For the operator W to be hyperbolic in S', it is necessary that the following conditions hold:

   (a) If λ is a multiplier of $V_0(\lambda, \sigma)$, then λ is also a multiplier of no less multiplicity of all the polynomials $V_s(\lambda, \sigma)$; $s = 1, 2. . .$,
   (b) $k_j \geq 0$ for all j such that $c_j \neq 0$.

2. Let the condition (a) be valid, and suppose that

(b') $k_j > 0$ for all j such that $c_j \neq 0$. Then the operator W is hyperbolic in S', $\operatorname{supp} E(t, x) \subseteq K$ and $E(t, x) \not\equiv 0$ in a however small neighbourhood of an arbitrary point $P \in \partial K \cap \partial \circ K$. Here $E(t, x)$ is the operator W fundamental solution describing finite speed wave propagation, K and $\circ K$ are the influence cone and the propagation cone for the operator V_0, respectively.

Corollary Suppose the relaxation kernel $R(t)$ can be represented in the form

$$R(t) = \phi(t) + a_1 \phi(t) * \phi(t) + \cdots$$

where $|a_i|$ are increasing not so rapidly as a geometrical progression. Then the wave operator with memory

$$\frac{\partial^2}{\partial t^2} - c^2 \left(\frac{\partial^2}{\partial x^2} - R(t) * \frac{\partial^2}{\partial x^2} \right)$$

is hyperbolic in S'. In fact, for this operator

$$k_{1,2} = \left. \frac{c^2 \sigma}{\lambda \frac{\partial}{\partial \lambda} \left(\lambda^2 - c^2 \sigma^2 \right)} \right|_{\lambda = \pm c\sigma, \sigma \neq 0} = \frac{1}{2} > 0.$$

Theorem 2.2.4 Let V_0 be strictly hyperbolic, and suppose

$$\varlimsup_{t \to +0} \frac{\phi(t)}{\ln(1/t)} = +\infty.$$

Then all the assertions of the previous theorem hold true for the space D'.

Proofs of these theorems are similar to the proofs of the corresponding assertions for operators containing a single convolution (see Theorems 2.1.1 and 2.1.3). Conditions of hyperbolicity given in Theorems 2.2.3 and 2.2.4 are close to necessary and sufficient ones. It is possible to specify these results in dependence on the character of growth of $\phi(t)$ as $t \to +0$.

2.3 Multidimensional Case: Preliminary Lemma

This section is devoted to geometrical constructions of which we shall have heed below when using the Paley–Wiener type theorem about analyticity in a tube domain (Theorem 1.3.2).

We consider λ and $\sigma = (\sigma_1, \ldots, \sigma_n)$ coordinates in $\mathbb{R}^1 \times \mathbb{R}^n$, $n > 1$.

Consider a cone

$$N = \{\lambda, \sigma | t\lambda + x.\sigma > 0 \text{ for } (t, x) \in K\}$$

where K is the influence cone for the strictly hyperbolic operator

$$V_0 = V_0 \left(\frac{\partial}{\partial t}, \frac{\partial}{\partial x_1}, \ldots, \frac{\partial}{\partial x_n} \right).$$

As we have noted in Sect. 1.2

$$N = \circ N \cap \{\lambda, \sigma | \lambda > 0\} \tag{2.3.1}$$

Here $\circ N$ denotes the core of the normal cone for the operator V_0.

Let us make a rotation in the λ, σ-space

$$g : \lambda, \sigma \to \lambda', \sigma' \tag{2.3.2}$$

under the condition that the semi-axis $\lambda' > 0$ belongs to the interior of the cone N. By virtue of (2.3.1), the last condition is equivalent to two following ones:

(a) The semi-axis $\lambda' > 0$ belongs to the interior of $\circ N$.
(b) The angle between semi-axes $\lambda > 0$ and $\lambda' > 0$ is acute.

The set of all rotations satisfying the above conditions, (a) and (b), will be denoted by G. A compact subset of G, consisting of all rotations under which the minimal (acute) angle between the semi-axis $\lambda' > 0$ and the surface of the cone N is greater than or equal to $\varepsilon > 0$, will be denoted by G_ε.

Let Ω be a sphere in $\mathbb{R}^n$ given by the equation $|\sigma| = 1$. Let us draw through an arbitrary vector $w \in \Omega$ and the λ'-axis a two-dimensional plane $\pi(w, g)$. Furthermore, let ρ' be the axis which lies in $\pi(w, g)$ and is orthogonal to the λ'-axis, and let w' be the unit directing vector of the ρ'-axis (supposing the angle between w and w' is acute). Clearly, when w runs over the sphere Ω, the vector w' runs over a unit sphere Ω', which is located in the hyperplane $\lambda' = 0$ and whose centre coincides with the origin of coordinates. Let us make the following rotation in $\pi(w, g)$

$$A(w, g) : \lambda', \rho' \to \lambda'', \rho'' \tag{2.3.3}$$

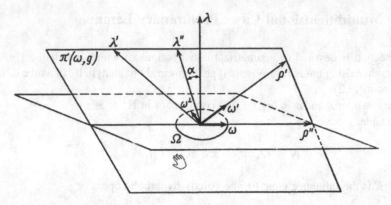

Fig. 2.2 A multidimensional case

where the λ''-axis is the projection of the λ-axis on $\pi(w, g)$ and the ρ''-axis is the intersection of $\pi(w, g)$ with the hyperplane $\lambda = 0$ (we define the positive direction on the λ''-axis as the projection of the positive direction on the λ-axis and consider w the directing vector of the ρ''-axis). Namely,

$$\lambda''(\lambda', \rho') = \lambda(\lambda', w'\rho') \frac{1}{\cos \alpha}, \tag{2.3.4}$$

$$\rho''(\lambda', \rho') = w \cdot \sigma(\lambda', w'\rho') \tag{2.3.5}$$

where $\alpha = \alpha(w, g)$ is the acute angle between the plane $\pi(w, g)$ and the λ-axis (see Fig. 2.2).

It is easy to see that

$$\sigma[\lambda'(0, \rho''), w'\rho'(0, \rho'')] = w\rho'', \tag{2.3.6}$$

$$\sigma[\lambda'(\lambda'', 0), w'\rho'(\lambda'', 0)] = w^{\perp}\lambda'' \sin \alpha \tag{2.3.7}$$

where $w^{\perp}$ is the unit directing vector of the projection of the λ''-axis on the hyperplane $\lambda = 0$. From (2.3.4, 2.3.6, and 2.3.7), it evidently follows that

$$\lambda[(\lambda'(\lambda'', \rho''), w'\rho'(\lambda'', \rho'')] = \lambda'' \cos \alpha, \tag{2.3.8}$$

$$\sigma[\lambda'(\lambda'', \rho''), w'\rho'(\lambda'', \rho'')] = w^{\perp}\lambda'' \sin \alpha + w\rho''. \tag{2.3.9}$$

Let us introduce into consideration the following quantities:

$$k_j(w) = \frac{V_1(\lambda, w\rho)}{\lambda \frac{\partial}{\partial \lambda} V_0(\lambda, w\rho)}\Bigg|_{\lambda = c_j(w)\rho, \rho \neq 0} \tag{2.3.10}$$

where $\rho \in \mathbb{R}^1$, $c_j(w)$ are coefficient of the decomposition

$$V_0(\lambda, w\rho) = V_0(1,0) \prod_{j=1}^{m} [\lambda - c_j(w)\rho]; \quad w \in \Omega \tag{2.3.11}$$

From strict hyperbolicity of the operator V_0, it follows that for $w \in \Omega$, the coefficients $k_j(w)$ are finite for w such that corresponding $c_j(w) \neq 0$. If for some $w \in \Omega$ there exists a $c_j(w) = 0$, then for w under consideration we shall say the corresponding coefficient $k_j(w)$ to be not determined. Let

$$\Omega^0 = \{w | V_0(0, w) = 0\}.$$

Then, on the set Ω^0, the number of determined coefficients $k_j(w)$ equals $m - 1$ (since, on account of strict hyperbolicity of V_0, exactly one of the coefficients $c_j(w)$ vanishes at each point of Ω^0). On the set $\Omega\backslash\Omega^0$, the number of determined coefficients k_j obviously equals m.

Furthermore, let us put

$$V_s^{w,g}(\lambda'', \rho'') = V_s\{\lambda[\lambda'(\lambda'', \rho''), w'\rho'(\lambda'', \rho'')], \sigma[\lambda'(\lambda'', \rho''), w'\rho'(\lambda'', \rho'')]\} \tag{2.3.12}$$

$$k_j(w, g) = \frac{V_1^{w,g}(\lambda'', \rho'')}{\lambda'' \frac{\partial}{\partial \lambda''} V_0^{w,g}(\lambda'', \rho'')}\Bigg|_{\lambda'' = c_j(w,g)\rho'', \rho'' \neq 0} \tag{2.3.13}$$

where $c_j(w)$ are coefficient of the decomposition

$$V_0^{w,g}(\lambda'', \rho'') = V_0^{w,g}(1,0) \prod_{j=1}^{m} [\lambda'' - c_j(w,g)\rho''] \tag{2.3.14}$$

Lemma 2.3.1 Let the operator V_0 be strictly hyperbolic. Then

(a) The polynomial $V_0[\lambda(\lambda', w'\rho'), \sigma(\lambda', w'\rho')]$ is strictly hyperbolic with respect to λ'.
(b) The polynomial $V_0^{w,g}(\lambda'', \rho'')$ is strictly hyperbolic with respect to λ''.
(c) The normal surface for the operator V_0 if bounded then in the λ'', ρ''-plane the intersection of the straight line $\lambda'' = 0$ with the cone $\{\lambda'', \rho'' | V_0^{w,g}(\lambda'', \rho'') = 0\}$ consist of the origin of coordinates.
(d) The semi-axis $\lambda' > 0$ belongs to the core of the cone $\{\lambda'', \rho'' | V_0^{w,g}(\lambda'', \rho'') = 0\}$; besides that, the angle between the semi-axes $\lambda' > 0$ and $\lambda'' > 0$ is acute.
(e) All the coefficients $c_j(w, g)$ are different.

(f) Let us fix an arbitrary $w \in \Omega$ and enumerate each of two sets of numbers $c_j(w)$;
$j = 1,\ldots, m$ and $c_j(w, g); j = 1,\ldots, m$ in order of increase. Then $c_{j_0}(w) = 0$ if and
only if $c_{j_0}(w, g) = 0$.

(g) Let $k_{j_0} > 0$ for w such that $c_{j_0}(w) \neq 0$ (here $w \in \Omega$). Then also $k_{j_0}(w, g) > 0$
for the same w.

Proof

(a) In accordance with our construction, the semi-axis $\lambda' > 0$ belongs to the core of
the cone $\{\lambda, \sigma | V_0(\lambda, \sigma) = 0\}$, whence the required assertion follows.

(b) Since the core of the cone $\{\lambda, \sigma | V_0(\lambda, \sigma) = 0\}$ is convex, the semi-axis $\lambda'' > 0$
(which is the orthogonal projection of the semi-axis $\lambda > 0$ on the plane $\pi(w, g)$) is
contained inside of the core of the cone $\{\lambda', \rho' | V_0[\lambda(\lambda', w'\rho'), \sigma(\lambda', w'\rho')] = 0\}$,
whence strict hyperbolicity of the polynomial $V_0^{w,g}(\lambda'', \rho'')$ follows. In particular,
$V_0^{w,g}(1, 0) \neq 0$.

(c) From (2.3.12, 2.3.8, 2.3.9), it follows that

$$V_0^{w,g}(0, \rho'') = V_0(0, w\rho'') \qquad (2.3.15)$$

Whence, by virtue of the assumption of boundedness of the normal surface for the
V_0, the required result follows.

(d) The fact that the semi-axis $\lambda' > 0$ is contained in the core of the cone
$\{\lambda'', \rho'' | V_0^{w,g}(\lambda'', \rho'') = 0\}$ follows from the point (b). The angle between the
semi-axes $\lambda' > 0$ and $\lambda'' > 0$ is acute, since it is evidently no greater than the
acute angle between the semi-axes $\lambda > 0$ and $\lambda' > 0$ (see Fig. 2.2).

(e) The assertion of this point follows from the point (b).

(f) Let us connect the mapping g with the identity mapping I by a continuous way of
mappings $g(Z) \in G, Z \in [0, 1]$, providing $g(0) = I, g(1) = g$. For each fixed
$w \in \Omega$, all numbers $c_j(w, g(Z))$ are different, continuously depend on Z, and
satisfy the condition $c_j(w, g(0)) = c_j(w); j = 1, \ldots, m$.

Furthermore, analogously to (2.3.15), we have

$$V_0^{w,g(Z)}(0, \rho'') = V_0(0, w\rho'')$$

whence

$$V_0^{w,g(Z)}(1, 0) \prod_{j=1}^{m} c_j(w, g(Z)) = V_0(1, 0) \prod_{j=1}^{m} c_j(w).$$

Therefore, for each $Z \in [0, 1]$ and each $w \in \Omega$, there exists j_0 such that
$c_{j_0}(w, g(Z)) = 0$ if and only if $c_{j_0}(w) = 0$. Now, the required assertion is proved.
Consider the expression

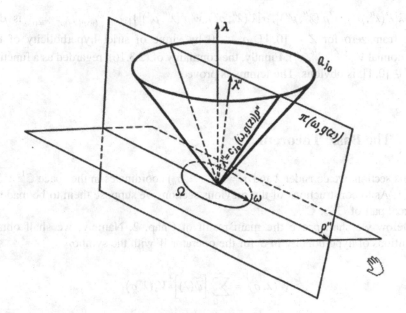

Fig. 2.3 Figure showing an acute angle between the semi-axes

$$k_{j_0}\left(w, g(Z)\right) = \frac{V_1^{w,g(Z)}(\lambda'', \rho'')}{\lambda'' \frac{\partial}{\partial \lambda''} V_0^{w,g(Z)}(\lambda'', \rho'')}\Bigg|_{\lambda''=c_{j_0}(w,g(Z))\rho'', \rho'' \neq 0} \tag{2.3.16}$$

for $Z \in [0, 1]$ under the condition $c_{j_0}(w) \neq 0$ (or, which is the same, $c_{j_0}(w, g(Z))$ / $= 0$). It is clear that

$$k_{j_0}(w, g(0)) = k_{j_0}(w); \quad k_{j_0}(w, g(1)) = k_{j_0}(w, g).$$

The required assertion will be established, if we demonstrate that for $Z \in [0, 1], c_{j_0}(w) \neq 0$, expression (2.3.16) is a non-vanishing continuous function of Z. (Without loss of generality, we consider $\rho'' > 0$ in (2.3.16).)

Notice, first of all, that from positiveness of $k_{j_0}(w)$ for $c_{j_0} \neq 0$ it follows that the polynomial $V_1(\lambda, \sigma)$ is distinct from zero on the cone

$$Q_{j_0} = \bigcup_{w \in \Omega, c_{j_0}(w) \neq 0} \{\lambda, \sigma | \lambda = c_{j_0}(w)\rho, \sigma = w\rho, \rho > 0\}. \tag{2.3.17}$$

Furthermore, it is geometrically evident that for $Z \in [0, 1], c_{j_0}(w) \neq 0$ the half-line

$$\lambda'' = c_{j_0}(w, g(Z))\rho'', \quad \rho'' > 0$$

belongs to the intersection of the plane $\pi(w, g(Z))$ with the cone (2.3.17) (see Fig. 2.3). Hence the fraction (2.3.16) numerator, which is equal to

$V_1\left\{\lambda[\lambda'(\lambda'',\rho''),w'\rho'(\lambda'',\rho'')],\sigma[\lambda'(\lambda'',\rho''),w'\rho'(\lambda'',\rho'')]\right\}|_{\lambda''=c_{j_0}(w,g(Z))\rho'',\rho''\neq 0}$ is distinct from zero for $Z \in [0,1]$, $\rho'' > 0$ by virtue of strict hyperbolicity of the polynomial $V_0^{w,g(Z)}(\lambda'',\rho'')$. Finally, the continuity of (2.3.16), regarded as a function of $Z \in [0,1]$, is obvious. The lemma is proved.

2.4 The Basic Theorem

In this section, we consider λ and $\sigma = (\sigma_1,\ldots,\sigma_n)$ coordinates in the space $\mathbb{C}^1 \times \mathbb{C}^n$, $n > 1$. As to constructions of the previous section, we suppose them to be made in the real part of $\mathbb{C}^1 \times \mathbb{C}^n$.

Below we shall prove the main result of Chap. 2. Namely, we shall obtain conditions of hyperbolicity in S' for the operator W with the symbol

$$W(\lambda,\sigma) = \sum_{s=0}^{\infty} \left[\widetilde{\phi}(\lambda)\right]^s V_s(\lambda,\sigma)$$

providing V_0 is a strictly hyperbolic operator with a bounded normal surface. From our assumptions it follows, in particular, that the numbers $c_j(w), j = 1,\ldots,m$, defined in (2.3.11) are all different and distinct from zero and that the coefficients

$$k_j(w) = \frac{V_1(\lambda,w\rho)}{\lambda\frac{\partial}{\partial\lambda}V_0(\lambda,w\rho)}\Bigg|_{\lambda=c_j(w)\rho,\rho\neq 0} \qquad ; \quad j = 1,\ldots,m$$

are determined for each $w \in \Omega$, where Ω is the unit sphere $|\text{Re}\sigma| = 1$ in the space $\text{Re}\sigma$.

Theorem 2.4.1 Let V_0 be a strictly hyperbolic operator with a bounded normal surface. Then the operator W is hyperbolic in S' and preserves this property under small real arbitrary perturbations of the coefficients of operators V_s; $s = 0, 1, \ldots$ if and only if

$$k_j(w) > 0; j = 1,\ldots,m \tag{2.4.1}$$

for each $w \in \Omega$.

The condition (2.4.1) if satisfied, $\text{supp}E(t,x) \subseteq \circ K$ and $E(t,x) \not\equiv 0$ in a however small neighbourhood of an arbitrary point $P \in \partial\circ K$. Here $E(t, x)$ is the operator W fundamental solution describing finite speed wave propagation, $\circ K$ is the propagation cone for the operator V_0.

Note We have already mentioned that in the multidimensional case $(n > 1)$, boundedness of the normal surface for V_0 yields the coincidence of the propagation cone for V_0 with the influence cone for V_0. This circumstance enables us to describe supp $E(t, x)$ more precisely than in case where $K \neq \circ K$.

Proof of the Theorem For simplicity, we shall, as usual, suppose $V_s(\lambda, \sigma) \equiv 0$ for $s \geq 3$; transition to the general case is obvious.

Necessity Suppose the operator W has a fundamental solution $E(t, x)$ describing finite speed wave propagation and such that $e^{-Mt} E(t, x) \in S'$ for some M. Let us prove the validity of (2.4.1), for example, for $w = (1, 0, \ldots, 0)$. Consider the function

$$u(t, x_1) = \int_{\mathbb{R}^{n-1}} E(t, x_1, x_2 - \xi_2, \ldots, x_n - \xi_n) \, d\xi_2 \ldots d\xi_n$$

On the one hand, it is easy to see that

$$\text{supp } u(t, x_1) \subseteq \{t, x_1 | t \geq \text{const } |x_1|\}, \quad \text{const} > 0,$$

and that $e^{-Mt} u(t, x_1) \in S'$. On the other hand, notice that the function $u(t, x_1)$ satisfies the equation

$$V_0\left(\frac{\partial}{\partial t}, \frac{\partial}{\partial x_1}, 0, \ldots, 0\right) u(t, x_1) + \phi(t) * V_1\left(\frac{\partial}{\partial t}, \frac{\partial}{\partial x_1}, 0, \ldots, 0\right) u(t, x_1)$$

$$+\phi(t) * \phi(t) * V_2\left(\frac{\partial}{\partial t}, \frac{\partial}{\partial x_1}, 0, \ldots, 0\right) u(t, x_1) = \delta(t)\delta(x_1) \tag{2.4.2}$$

Hence the operator on the left-hand side of (2.4.2) must satisfy the necessary conditions of Theorem 2.2.3. Therefore, taking into account the requirement of stability, we obtain the inequalities (2.4.1) for $w = (1, 0, \ldots, 0)$. The case of arbitrary w obviously reduces to the case of $w(1, 0, \ldots, 0)$ by means of a rotation.

Sufficiency We are going to show that the function

$$\frac{1}{i^m W(\lambda, \sigma)}$$

is a Fourier–Laplace transform of a distribution $E(t, x)$ such that $e^{-Mt} E(t, x) \in S'$ (where $M > 0$ is large enough) and supp $E(t, x) \subseteq \circ K$, whence it will evidently follow that $E(t, x)$ is the desired fundamental solution for the operator W.

Now, note the following circumstance. Real homogeneous linear mappings (2.3.2) and (2.3.3) can be naturally extended to the complex space. Under such an extension, the imaginary parts of variables will be transformed by the same law as the real ones. For the extended, in such a way, mappings (2.3.2, 2.3.3), we shall use the former notation. If we demonstrate that under conditions:

(a) $g \in G_\varepsilon$ (where $\varepsilon > 0$ is sufficiently small)

(b) $\text{Im}\lambda' \leq -B$ (where $B = B(\varepsilon) > 0$ is sufficiently large);$\sigma' \in \mathbb{R}^n$ the estimate

$$\left| W\left(\lambda(\lambda', \sigma'), \sigma(\lambda', \sigma') \right) \right|^{-1} \leq \text{const} \tag{2.4.3}$$

holds true, then by Theorem 1.4.2, we shall obtain the existence of $E(t, x)$ with the required properties. Let $w \in \Omega$, $g \in G_\varepsilon$. Consider the expression

$$W^{w,g}(\lambda'', \rho'') = V_0^{w,g}(\lambda'', \rho'') +$$
$$+ \left[\widetilde{\phi}(\lambda \cos \alpha) \right]^2 V_2^{w,g}(\lambda'', \rho'')$$

where $\alpha = \alpha(w, g)$ is the acute angle between the axis $\text{Re}\lambda''$ and the plane$\pi(w, g)$. We recall that by virtue of Lemma 2.3.1 (e),(f), all the numbers $c_j(w, g)$ are different and distinct from zero and all the coefficients $k_j(w, g)$ are determined. Hence from Lemma 2.2.1, for continuity and compactness reasons, it follows that for $-\text{Im }\lambda''$ large enough, the roots $\rho'' = \rho_j(\lambda'', w, g)$ of the equation

$$W^{w,g}(\lambda'', \rho'') = 0$$

satisfy   the   following   relations $c_j(w, g) \text{Im}\rho_j(\lambda'', w, g) = \text{Im}\lambda'' +$
$\left[k_j(w, g) + o(1) \right] \text{Im} \left[\lambda'' \widetilde{\phi}(\lambda'' \cos \alpha) \right]$; for $j = 1, \ldots, m$, where $o(1)$ goes to zero uniformly with respect to $\text{Re}\lambda''$, w, g, as $\text{Im}\lambda'' \to -\infty$. Furthermore, from (2.4.1) and Lemma 2.3.1(g), we easily obtain, for continuity and compactness reasons, that

$$k_j(w, g) \geq k(\varepsilon) > 0; \quad j = 1, \ldots, m$$

Therefore, by virtue of Lemma 1.5.4

$$c_j(w, g) \, \text{Im}\rho_j(\lambda'', w, g) \leq \text{Im}\lambda'', \quad \text{for} \, \text{Im}\lambda'' \leq -B_1$$

where $B_1 = B_1(\varepsilon)$ is large enough. Therefore

$$\left| W^{w,g}(\lambda'', \rho'') \right|^{-1}$$
$$= \left| \left\{ V_0^{w,g}(0, 1) + \widetilde{\phi}(\lambda'' \cos \alpha) V_1^{w,g}(0, 1) + \left[\widetilde{\phi}(\lambda'' \cos \alpha) \right]^2 V_2^{w,g}(0, 1) \right\} \right|$$
$$\times \left| \prod_{j=1}^{m} \left[\rho'' - \rho_j(\lambda'', w, g) \right]^{-1} \right| \leq \text{const} \tag{2.4.4}$$

on the set

$$T(w,g,\delta)=\{\lambda'',\rho''\,|\,\text{Im}\lambda'' \leq \min\left[(1+\delta)c_1(w,g)\text{Im}\rho'',\ldots\ldots,(1+\delta)c_m(w,g)\text{Im}\rho'',-B_1\right]\},$$

$$(2.4.5)$$

where $\delta > 0$ is whatever small.

Furthermore, by virtue of Lemma 2.3.1(d), for fixed $w \in \Omega$, $g \in G_e$, one can choose $B > 0$ large enough and $\delta > 0$ small enough for the inclusion

$$\{\lambda',\rho'\,|\,\text{Im}\lambda' \leq -B, \rho' \in \mathbb{R}^1\} \subseteq T(w,g,\delta)$$

$$(2.4.6)$$

to be valid. Here λ', ρ' and λ'', ρ'' are related by the rotation $A(w,g)$. Moreover, by virtue of continuous dependence of $A(w,g)$ on parameters w and g, one can always consider $B = B(\varepsilon) > 0$ and $\delta = \delta(\varepsilon) > 0$ to be such that (2.4.6) holds true for all $w \in \Omega$, $g \in G_e$. Now, from (2.4.4, 2.4.5, and 2.4.6), it follows that

$$\left|W^{w,g}\left(\lambda''(\lambda',\rho'),\rho''(\lambda',\rho')\right)|^{-1} \leq \text{const}\right.$$

$$(2.4.7)$$

for $\text{Im}\lambda' \leq -B$, $\rho' \in \mathbb{R}^1$, $w \in \Omega$, $g \in G_e$..However

$$W^{w,g}(\lambda''(\lambda',\rho'),\rho''(\lambda',\rho')) \equiv W(\lambda(\lambda',w'\rho'),\sigma(\lambda',w'\rho')),$$

hence (2.4.7) is equivalent to the desired inequality (2.4.3). Therefore, the operator W has a fundamental solution $E(t, x)$ such that supp $E(t,x) \subseteq \circ K$ and $e^{-Mt}E(t, x) \in S'$ for some $M > 0$ (one can assume M to be equal to $B(\varepsilon_0)$, where $\varepsilon_0 > 0$ is a sufficiently small fixed number).

Now, let P be an arbitrary point of $\partial \circ K$. Let us demonstrate that $E(t, x)$ does not identically vanish in a however small neighbourhood of the point P.

Suppose the Contrary Let $E(t, x) \equiv 0$ on the open set U containing the point P. Let us draw, through P, a hyperplane tangent to the cone $\circ K$ and choose coordinate axes $x_2, \ldots, x_n$ in the intersection of hyperplanes π_p and $t = 0$; the x_1- axis we direct along the normal to the mentioned above intersection. (The origin of coordinates is, as before, located at the vertex of the cone $\circ K$). Then the projection of supp $E(t, x)$ on the t, x_1-plane will not contain some neighbourhood U_1 of the point P_1 (P_1 is the projection of the point P on the t, x_1-plane).

Let

$$u(t,x_1) = \int_{\mathbb{R}^{n-1}} E(t,x_1,x_2 - \xi_2, \ldots,x_n - \xi_n)\,d\xi_2 \ldots d\xi_n.$$

It is obvious that the function $u(t,x_1)$ identically vanishes on the set U_1 defined above.

Fig. 2.4 Figure showing
the position of point P_1

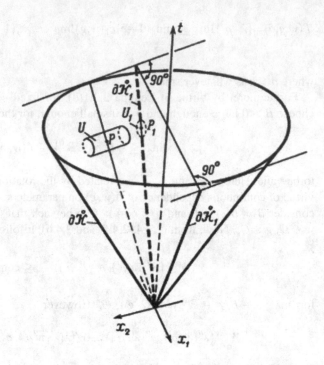

As was noted in the proof of necessity, the function $u(t, x_1)$ satisfies the equation

$$V_0\left(\frac{\partial}{\partial t}, \frac{\partial}{\partial x_1}, 0, \ldots, 0\right) u(t, x_1) + \phi(t) * V_1\left(\frac{\partial}{\partial t}, \frac{\partial}{\partial x_1}, 0, \ldots, 0\right) u(t, x_1) + \phi(t)$$

$$* \phi(t) * V_2\left(\frac{\partial}{\partial t}, \frac{\partial}{\partial x_1}, 0, \cdots, 0\right) u(t, x_1)$$

$$= \delta(t)\delta(x_1)$$

Besides that, it is geometrically evident that the influence cone for the operator $V_0(\partial/\partial t, \partial/\partial x_1, 0, \ldots, 0)$ coincides with the propagation cone $\circ K_1$ for this operator. Therefore by virtue of Theorem 2.2.3, supp $u(t, x_1) \subseteq \circ K_1$ and $u(t, x_1)$ does not identically vanish in a however small neighbourhood of an arbitrary point on $\partial \circ K_1$.

However, it is easy to see that the cone $\circ K_1$ is just the projection of the cone $\circ K$ on the t, x_1- plane. It is also clear that the point P_1 belongs to $\partial \circ K_1$ (see Fig. 2.4).

Hence the function $u(t, x_1)$ cannot identically vanish in the neighbourhood of the point P_1, which gives the desired contradiction.

Corollary Suppose the relaxation kernel R(t) can be represented in the form

$$R(t) = \phi(t) + a_1 \phi(t) * \phi(t) + \cdots$$

where $|a_i|$ is increasing not so rapidly as a geometrical progression. Then the wave operator with memory

$$\frac{\partial^2}{\partial t^2} - c^2 \left[\sum_{k=1}^{n} \frac{\partial^2}{\partial x_k^2} - R(t) * \sum_{k=1}^{n} \frac{\partial^2}{\partial x_k^2} \right]$$

is hyperbolic in S'. In fact, for this operator

$$k_{1,2}(w) = \frac{c^2 \rho^2}{\lambda \frac{\partial}{\partial \lambda} \left(\lambda^2 - c^2 \rho^2 \right)} \bigg|_{\lambda = \pm c\rho, \rho \neq 0} = \frac{1}{2} > 0.$$

2.5 Generalization of the Basic Theorem to the Case of Unbounded Normal Surface

In the multidimensional case, unboundedness of the normal surface for the operator V_0 leads us to difficulties, analogous to the ones which occur for hyperbolic differential operators with characteristics of variable multiplicity [7]. In what follows we shall use the former notation.

Theorem 2.5.1 Let V_0 be a strictly hyperbolic operator, Then the operator W is hyperbolic in S' only if for each $w \in \Omega$

(a) $\{w| V_0(0, w) = 0\} \subseteq \{w| V_s(0, w) = 0\}$; $s = 1, 2, \ldots,$
(b) $k_j(w) \geq 0$ for all j such that $c_j(w) \neq 0$.

To prove this theorem, it suffices to suppose the contrary and make use of the Hadamard's method of descent, which leads us to the desired contradiction with the results obtained above in the one-dimensional case (see Theorem 2.2.3).

Theorem 2.5.2 Let V_0 be a strictly hyperbolic operator and let for each $w \in \Omega$

(a) $\{w| V_0(0, w) = 0\} \subseteq \{w| V_s(0, w) = 0\}$; $s = 1, 2, \ldots,$
(b) $\sum_{i=1}^{n} \left[\frac{\partial}{\partial w_i} V_0(0, w) \right]^2 \bigg|_{V_0(0,w)=0} \neq 0,$

(c) $\displaystyle\sum_{i=1}^{n} \left[\frac{\partial}{\partial w_i} \prod_{j=1}^{m} V_1\big(c_j(w), w\big) \right]^2 \Bigg|_{V_0(0,w)=0} \neq 0,$

(d) $k_j(w) > 0$ for all j such that $c_j(w) \neq 0$. Then the operator W is hyperbolic in S', $\operatorname{supp} E(t,x) \subseteq K$ and $E(t,x) \not\equiv 0$ in a however small neighbourhood of arbitrary point $P \in \partial K \cap \partial \circ K$. Here $E(t, x)$ is the operator W fundamental solution describing finite speed wave propagation; K and $\circ K$ are the influence cone and the propagation cone for the operator V_0, respectively.

The proof of this theorem can be carried out in the manner of the proof of sufficiency in Theorem 2.4.1. Here we shall only show the changes which are to be introduced into the proof of the inequality (2.4.3) as case of unbounded normal surface for the operator V_0; for simplicity we suppose g to be the identity mapping (the general case can be studied in the same manner). Namely, we shall demonstrate that for $\operatorname{Im}\lambda \leq -M$, where $M > 0$ is large enough, $\sigma \in \mathbb{R}^n$, the inequality

$$\frac{1}{|W(\lambda, \sigma)|} \leq \text{const} \tag{2.5.1}$$

holds true.

In the neighbourhood of points $w \in \Omega$ such that $V_0(0, w) \neq 0$, we have in fact established (2.5.1) in the proof of Theorem 2.4.1. Now, let us establish the validity of (2.5.1) in some neighbourhood of the set

$$\Omega^0 = \{w | V_0(0, w) = 0\}$$

Let us put $\sigma = w\rho$ and decompose the symbol of the operator V_0 into a product of first-order multipliers

$$V_0(\lambda, w\rho) = V_0(1, 0) \prod_{j=1}^{m} \big(\lambda - c_j(w)\rho\big)$$

(We suppose the coefficients $c_j(w), j = 1, .., m$ to be enumerated in order of increase.) Let us fix an arbitrary $w^0 \in \Omega^0$ and suppose, to be specific, $c_1(w^0) = 0$.

Furthermore, let us expand $V_s(\lambda, w\rho)$ in powers of $\lambda - c_1(w)\rho$, then the characteristic equation will take the following form

$$
\begin{aligned}
W(\lambda, w\rho) \equiv {}& [\lambda - c_1(w)\rho]\{a_{m-1}(w)[\lambda - c_1(w)\rho]^{m-1} \\
&+ a_{m-2}(w)[\lambda - c_1(w)\rho]^{m-2}\rho \\
&+ \cdots + a_0(w)\rho^{m-1}\} \\
&+ \widetilde{\phi}(\lambda)\{b_m(w)[\lambda - c_1(w)\rho]^m + b_{m-1}(w)[\lambda - c_1(w)\rho]^{m-1}\rho \quad (2.5.2) \\
&+ \cdots + b_0(w)\rho^m\} \\
&+ \left[\widetilde{\phi}(\lambda)\right]^2 \{d_m(w)[\lambda - c_1(w)\rho]^m \\
&+ d_{m-1}(w)[\lambda - c_1(w)\rho]^{m-1}\rho + \cdots + d_0(w)\rho^m\} = 0
\end{aligned}
$$

For simplicity, we have supposed, as usual, $V_s(\lambda, \sigma) \equiv 0$ for $s \geq 3$.
  Letting

$$
Z = \frac{\lambda - c_1(w)\rho}{\rho\widetilde{\phi}(\lambda)} \tag{2.5.3}
$$

and inserting (2.5.3) into (2.5.2), we obtain by cancellation by $\rho^m\widetilde{\phi}(\lambda)$,

$$
Z\left[a_{m-1}(w)\left(\widetilde{\phi}(\lambda)\right)^{m-1}Z^{m-1} + \cdots + a_0(w)\right] + b_m\left(\widetilde{\phi}(\lambda)\right)^m Z^m + \ldots + b_0(w)
$$
$$
+ \widetilde{\phi}(\lambda)\left[d_m(w)\left(\widetilde{\phi}(\lambda)\right)^m Z^m + \ldots + d_0(w)\right] = 0
$$
$$\tag{2.5.4}$$

  By virtue of strict hyperbolicity of the operator V_0, it is evident that $a_0(w) \neq 0$ in some neighbourhood of the point w^0. Besides that, it follows from the condition (a) of the theorem that $b_0(w^0) = 0, d_0(w^0) = 0$. Therefore, by the theorem about implicit functions, it follows that for however small $\varepsilon > 0$, there do exist $M = M(\varepsilon) > 0$ and $\delta = \delta(\varepsilon) > 0$, such that for $\mathrm{Im}\lambda \leq -M, |w - w^0| \leq \delta$, Eq. (2.5.4) has a continuous solution $Z = Z_1(\lambda, w)$ satisfying the estimate

$$
|Z_1(\lambda, w)| \leq \varepsilon. \tag{2.5.5}
$$

  Now, assembling in (2.5.4) terms of first and second orders in Z and taking into account (2.5.5), we obtain that for $\mathrm{Im}\lambda \leq -M, |w - w^0| \leq \delta$,

$$
\left| Z_1(\lambda, w)\left\{a_0(w) + \widetilde{\phi}(\lambda)b_1(w) + \left[\widetilde{\phi}(\lambda)\right]^2 d_1(w)\right\} + b_0(w) + \widetilde{\phi}(\lambda)d_0(w) \right|
$$
$$
\leq \mathrm{const}\,|Z_1^2(\lambda, w)|
$$

where const > 0 does not depend on λ, w. Hence, as $\varepsilon \to 0$,

$$Z_1(\lambda, w) = -\frac{b_0(w) + \widetilde{\phi}(\lambda)\, d_0(w)}{a_0(w) + \widetilde{\phi}(\lambda)\, b_1(w) + \left[\widetilde{\phi}(\lambda)\right]^2 d_1(w)}$$

$$+ O\left(\left\{\frac{b_0(w) + \widetilde{\phi}(\lambda)\, d_0(w)}{a_0(w) + \widetilde{\phi}(\lambda)\, b_1(w) + \left[\widetilde{\phi}(\lambda)\right]^2 d_1(w)}\right\}^2\right) \tag{2.5.6}$$

uniformly with respect to λ, w for

$$\mathrm{Im}\lambda \le -M, \qquad |\, w - w^0\,| \le \delta.$$

One can directly calculate that for w such that $c_1(w) \ne 0$ the equality

$$\frac{b_0(w)}{a_0(w)\, c_1(w)} = k_1(w) \tag{2.5.7}$$

holds. Furthermore, from the equality

$$V_0(0, w) = (-1)^m\, V_0(1, 0) \prod_{j=1}^{m} c_j(w)$$

and condition (b) of the theorem, it follows that

$$\sum_{i=1}^{n} \left[\frac{\partial c_1(w)}{\partial w_i}\right]^2\Bigg|_{c_1(w)=0} \ne 0. \tag{2.5.8}$$

Besides that, the equality $V_1(c_1(w), w) = b_0(w)$ and condition (c) of the theorem yield

$$\sum_{i=1}^{n} \left[\frac{\partial b_0(w)}{\partial w_i}\right]^2\Bigg|_{c_1(w)=0} \ne 0, \tag{2.5.9}$$

since, by virtue of condition (a) of the theorem,

$$b_0(w)\big|_{c_1(w)=0} = 0. \tag{2.5.10}$$

From (2.5.8, 2.5.9, and 2.5.10), it easily follows that the function $k_1(w)$ can be extended by continuity to the set of those w for which $c_1(w) = 0$. Moreover, it is evident that the inequality $k_1(w) > 0$, which, by condition (d), is valid for w such that $c_1(w) \ne 0$, will hold true under such an extension. From continuity of the (extended) function $k_1(w)$ and compactness of the sphere Ω, it follows that a stronger inequality holds

$$k_1(w) \geq \text{const} > 0; \quad w \in \Omega. \tag{2.5.11}$$

Since the function $k_1(w)$ is now determined for all $w \in \Omega$, it follows from (2.5.6, 2.5.7) that for $\text{Im}\lambda \leq -M$, $|w - w^0| \leq \delta$ the following ratio is also determined:

$$\frac{Z_1(\lambda, w)}{c_1(w)} = -k_1(w) + o(1) + O(|w - w^0|). \tag{2.5.12}$$

Here the quantity $o(1)$ tends to zero uniformly with respect to $\text{Re}\lambda$, w as $\text{Im}\lambda \to -\infty$; the quantity $O(|w - w^0|)$ is uniform with respect to λ.

The root $Z_1(\lambda, w)$ of (2.5.4) evidently corresponds to the root $\rho_1(\lambda, w)$ of (2.5.2) such that

$$c_1(w)\rho_1(\lambda, w) = \frac{\lambda c_1(w)}{c_1(w) + Z_1(\lambda, w)\widetilde{\phi}(\lambda)}$$

$$= \frac{\lambda}{1 - [k_1(w) + o(1) + O(|w - w^0|)]\widetilde{\phi}(\lambda)}$$

$$= \lambda + [k_1(w) + o(1) + O(|w - w^0|)]\lambda\widetilde{\phi}(\lambda).$$

By virtue of Lemma 1.5.5, the last relation yields

$$c_1(w)\,\text{Im}\rho_1(\lambda, w) = \text{Im}\lambda + [k_1(w) + o(1) + O(|w - w^0|)]\text{Im}\left[\lambda\widetilde{\phi}(\lambda)\right]. \tag{2.5.13}$$

In what follows, we consider $M > 0$ sufficiently large and $\delta > 0$ sufficiently small. From (2.5.11, 2.5.13) and Lemma 1.5.4, it follows that

$$c_1(w)\,\text{Im}\rho_1(\lambda, w) \leq \text{Im}\lambda \quad \text{for}\,\text{Im}\lambda \leq -M, \ |w - w^0| \leq \delta \tag{2.5.14}$$

Furthermore, by virtue of strict hyperbolicity of the operator V_0, $c_j(w) \neq 0$ for $|w - w^0| \leq \delta; j = 2, \ldots, m$, whence on account of Lemma 2.2.1, it follows that for large $-\text{Im}\,\lambda$ and $|w - w^0| \leq \delta$, the characteristic equation $W(\lambda, \sigma) = 0$ has the roots $\rho = \rho_j(\lambda, w); j = 2, \ldots, m$ satisfying the relations

$$c_j(w)\,\text{Im}\rho_j(\lambda, w) = \text{Im}\lambda + [k_j(w) + o(1)]\,\text{Im}\left[\lambda\widetilde{\phi}(\lambda)\right]$$

Here the quantity $o(1)$ tends to zero uniformly with respect to $\text{Re}\lambda$, w as $\text{Im}\lambda \to -\infty$. Hence, in accordance with condition (d) and Lemma 1.5.4, we obtain the following inequalities analogous to (2.5.14)

$$c_j(w)\,\mathrm{Im}\rho_j(\lambda,w) \le \mathrm{Im}\lambda \text{ for } \mathrm{Im}\lambda \le -M, \left|w-w^0\right| \le \delta; \tag{2.5.15}$$

for $j = 2,\ldots, m$

Thus from (2.5.14, 2.5.15) and condition (a) of the theorem, we obtain that the function

$$\frac{1}{W(\lambda,w\rho)}$$

$$=\frac{\displaystyle\prod_{j=1}^{m} c_j(w)}{\left[V_0(0,w)+\widetilde{\phi}(\lambda)V_1(0,w)+\left(\widetilde{\phi}(\lambda)\right)^2 V_2(0,w)\right]\displaystyle\prod_{j=1}^{m}\left(c_j(w)\rho-c_j(w)\rho_j(\lambda,w)\right)}$$

$$=\frac{(-1)^m V_0(0,w)}{V_0(1,0)\left[V_0(0,w)+\widetilde{\phi}(\lambda)V_1(0,w)+\left[\widetilde{\phi}(\lambda)\right]^2 V_2(0,w)\right]\prod_{j=1}^{m}\left[c_j(w)\rho-c_j(w)\rho_j(\lambda,w)\right]}$$

is bounded for $\mathrm{Im}\lambda \le -M, \rho \in \mathbb{R}^1, \left| w - w^0 \right| \le \delta$.

Finally, covering Ω^0 with neighbourhoods of the $\left|w - w^0\right| \le \delta$ type and choosing a finite sub-covering from this covering, we arrive at the desired inequality (2.5.1).

Note Chap. 2 is based on the results of [8, 9].

References

1. Vladimirov, V.S.: *Distributions in Mathematical Physics*, pp. 1–318. Nauka, Moscow (1979)
2. Marcushevitc, A.l.: *Theory of Analytic Functions*, vol. 1, pp. 1–486. Nauka, Moscow (1967)
3. Lions, J.-L. *J. Anal. Math.* 2, 369, 1952–1953.
4. Atiah, M.F., Bott, R., Garding, L.: *Acta Math.* **24**, 109 (1970)
5. Lax, A.: *Comm. Pure Appl. Math.* **9**, 135 (1956)
6. Münster, M. *Rocky Mountain J. of Math.* 3, 443 (1978).
7. Courant, R., Lax, A.: *Comm. Pure Appl. Math.* **8**, 497 (1955)
8. Lokshin, A.A., Suvorova, J.V.: *Mathematical Theory of Wave Propagation in Media with Memory*, pp. 1–151. Moscow University Press, Moscow (1982)
9. Lokshin, A.A.: *Trudy Sem. Petrovskogo*, vol. 7, p. 148 (1982)

Chapter 3
The Wave Equation with Memory

In this chapter, we study wave front asymptotics of solutions of wave equations with memory. Sections 3.1, 3.2, 3.3, 3.4, 3.5, 3.6, 3.7, 3.8, and 3.9 are devoted to the one-dimensional case. In Sects. 3.10 and 3.11, we deal with the case of two and three spatial variables, respectively.

3.1 Formulation of the Problem

Let us consider a linear hereditary elastic homogeneous rod located on the semi-axis $x \geq 0$. Then, as it easily follows from (1.1.9), the deformation in such a rod satisfies the following wave equation with memory

$$[1 + \Lambda(t)*]\frac{\partial^2 \varepsilon}{\partial t^2} - c^2 \frac{\partial^2 \varepsilon}{\partial x^2} = 0 \qquad (3.1.1)$$

Here $c = (A/\rho)^{1/2}$, $\rho = \text{const} > 0$ is the rod density, and $A = \text{const} > 0$ is the instantaneous module of elasticity.

Let us set the following problem for Eq. (3.1.1):

$$\varepsilon(t,x) = 0, \quad \text{for } x > 0, \ t \leq 0;$$

$$\varepsilon(t,0) = \begin{cases} 1 & \text{for } t > 0 \\ 0 & \text{for } t < 0 \end{cases} \qquad (3.1.2)$$

If we impose geometrical restrictions 1–5 from Sect. 1.1.8 on the kernel $\Lambda(t)$, then (3.1.1) will evidently describe the finite speed wave propagation. However, in this chapter, we shall use another approach to the problem (3.1.1, 3.1.2). Namely,

A. A. Lokshin, *Tauberian Theory of Wave Fronts in Linear Hereditary Elasticity*,
https://doi.org/10.1007/978-981-15-8578-4_3

following [1], along the whole of Chap. 3, we suppose that the creep kernel $\Lambda(t)$ can be represented as

$$\Lambda(t) = \varphi(t) + \frac{1}{4}\varphi(t) * \varphi(t) \tag{3.1.3}$$

where the function $\varphi(t)$ is supposed satisfying the following conditions:

$$\varphi(t) = 0 \text{ for } t < 0;$$

(a) $\varphi(t)$ is non-negative, non-increasing and concave for $t > 0$.
(b) $\varphi(t)$ is n-smooth for $t > 0$ and satisfies the estimate

$$\left|\frac{d^n \varphi(t)}{dt^n}\right| \leq \text{const } t^{-n-\delta}, \quad 0 < \delta < 1,$$

for small $t > 0$, where it is supposed that $n(1 - \delta) > 1/2$;

(c) $\varphi(t) \to +\infty$ as $t \to +0$, $\varphi(t) \to 0$ as $t \to +\infty$;
(d) $\varphi(t)$ is locally integrable on $[0, \infty]$.

By applying the Fourier–Laplace transform to (3.1.1, 3.1.2, and 3.1.3), one can easily obtain

$$\widetilde{\varepsilon}(\lambda, x) = \frac{e^{-i\lambda\sqrt{1+\widetilde{\Lambda}(\lambda)}\,(x/c)}}{i\lambda} = \frac{e^{-i\lambda\left\{1+\left[\widetilde{\varphi}(\lambda)/2\right]\right\}(x/c)}}{i\lambda} \tag{3.1.4}$$

where $\lambda = \mu - ip, p > 0$.

Note that from conditions (a) and (b), it follows that

$$\text{Im}\left[\lambda\widetilde{\varphi}(\lambda)\right] \leq 0, \text{ for } p > 0 \tag{3.1.5}$$

(See the proof of Lemma 1.5.4.) Therefore, for $p \geq \text{const} > 0$

$$\left|\widetilde{\varepsilon}(\lambda, x)\right| \leq \frac{\text{const}}{|\lambda|} e^{-px/c} \tag{3.1.6}$$

whence it follows that

$$\varepsilon(t, x) = 0, \text{ for } t < \frac{x}{c} \tag{3.1.7}$$

(See Theorem 1.3.1.)
Now, suppose that for some $x > 0$ and some small $\delta > 0$,

$$\varepsilon(t, x) = 0, \quad \text{for } t < \frac{x}{c - \delta}.$$

Then by Theorem 1.3.1, we should have for $p \geq M$, where $M > 0$ is large enough:

$$| \widetilde{\varepsilon}(\lambda, x) | \leq C(1 + |\lambda|)^{\upsilon} \, e^{-p[x/(c-\delta)]}; \tag{3.1.8}$$

where $C = \text{const} > 0$, $\upsilon = \text{const} > 0$.

Note, however, that putting in (3.1.4) $\lambda = -ip$, we have

$$\widetilde{\varepsilon}(-ip, x) = \frac{1}{p} \exp \left\{ -p \left[1 + \frac{\widetilde{\varphi}(-ip)}{2} \right] \frac{x}{c} \right\}. \tag{3.1.9}$$

By virtue of Lemma 1.5.1 $\widetilde{\varphi}(-ip) \to 0$ as $p \to \infty$, hence (3.1.9) contradicts (3.1.8).

Finally, note that by virtue of the estimate (3.1.6), the function $\widetilde{\varepsilon}(\lambda, x)$ is square integrable over $d\mu$ along each straight line $p = \gamma$, $\gamma > 0$, whence by Plancherel's theorem, it follows that $e^{-\gamma t}\varepsilon(t, x)$ is square integrable over dt on $[0, \infty)$.

Hence, in its turn, it follows that the function $e^{-\delta t}\varepsilon(t, x)$ is Lebesgue integrable over dt on $[0, \infty)$ for each $\delta > \gamma > 0$ (and hence for each $\delta > 0$). Therefore, for $p > 0$, one can consider the Fourier–Laplace transform $\widetilde{\varepsilon}(\lambda, x)$ in the classical sense.

Thus we have arrived at the following result.

Theorem 3.1.1 The solution $\varepsilon(t, x)$ of (3.1.1, 3.1.2, and 3.1.3) vanishes for $t < x/c$, does not identically vanish in a however small neighbourhood of an arbitrary point on the wave front $t = x/c$ and is locally integrable over dt. Moreover, the classical Laplace transform $L_{t \to p}\varepsilon = \overline{\varepsilon}(p, x)$ makes sense for $p > 0$.

In terms of the Laplace transform formula (3.1.9) evidently assumes the form

$$\overline{\varepsilon}(p, x) = \frac{1}{p} \exp \left\{ -p \left[1 + \frac{\overline{\varphi}(p)}{2} \right] \frac{x}{c} \right\} \tag{3.1.9'}$$

Thus using the well-known properties of the Fourier and Laplace transforms, one can write the following formula for the solution of (3.1.1, 3.1.2, and 3.1.3):

$$\varepsilon(t, x) = F^{-1}_{\lambda \to t - (x/c)} \frac{1}{i\lambda} \exp \left[-i\lambda \widetilde{\varphi}(\lambda) \frac{x}{2c} \right] = L^{-1}_{p \to t - (x/c)} \frac{1}{p} \exp \left[-\overline{\varphi}(p) \frac{x}{2c} \right] \tag{3.1.10}$$

Our main purpose is to give a description of the asymptotic behaviour of (3.1.10) in the vicinity of the wave front $t = x/c$ in dependence on the character of singularity of $\varphi(t)$. Lemma 3.1.1 demonstrates the fact that the sought for asymptotic behaviour of $\varepsilon(t, x)$ in the vicinity of $t = 0$ depends only on the behaviour of $\varphi(t)$ in the vicinity of $t = 0$.

Lemma 3.1.1 Let $\psi(t)$ be a locally integrable function equal to zero for $t < 0$, smooth for $t > 0$ and satisfying the estimate

$$| \psi'(t) | \le \text{const } e^{\gamma t}, \gamma > 0, \text{ for large } t > 0.$$

(a) Let

$$\psi(t) = \varphi(t), \text{ for } t < t_0, t_0 > 0 \qquad (3.1.11)$$

Then the function $\varepsilon(t, x)$ determined by (3.1.10) will not change in the domain $t < x/c + t_0$ if one substitutes φ for ψ in (3.1.10).

(b) Let $\varphi_1(t) = \varphi(t) + v \ln t_+$ where $v = \text{const}$ is a real number:

$$\ln t_+ = \begin{cases} \ln t & \text{for } t > 0 \\ 0 & \text{for } t < 0 \end{cases}$$

Then

$$F_{\lambda \to t}^{-1} e^{-i\lambda \widetilde{\varphi}_{11}(\lambda)} = 0, \quad \text{for } t < 0. \qquad (3.1.12)$$

Furthermore, let

$$\psi(t) = \varphi_1(t), \quad \text{for } t < t_0, \ t_0 > 0. \qquad (3.1.13)$$

Then

$$F_{\lambda \to t}^{-1} e^{-i\lambda \widetilde{\psi}(\lambda)} = F_{\lambda \to t}^{-1} e^{-i\lambda \widetilde{\varphi}_{11}(\lambda)}, \quad \text{for } t < t_0. \qquad (3.1.14)$$

Proof
(a) For $\lambda = \mu - ip, p > \gamma > 0$, we have

$$\frac{1}{i\lambda} \exp\left\{ -i\lambda \left[1 + \frac{\widetilde{\varphi}(\lambda)}{2} \right] \frac{x}{c} \right\} - \frac{1}{i\lambda} \exp\left\{ -i\lambda \left[1 + \frac{\widetilde{\psi}(\lambda)}{2} \right] \frac{x}{c} \right\}$$
$$= \frac{1}{i\lambda} \exp\left\{ -i\lambda \left[1 + \frac{\widetilde{\varphi}(\lambda)}{2} \right] \frac{x}{c} \right\} \left(1 - \exp\left\{ i\lambda [\widetilde{\varphi}(\lambda) - \widetilde{\psi}(\lambda)] \frac{x}{2c} \right\} \right) \qquad (3.1.15)$$

Note that by virtue of (3.1.11)

$$i\lambda[\widetilde{\varphi}(\lambda) - \widetilde{\psi}(\lambda)] = i\lambda \int\limits_{t_0}^{\infty} [\varphi(t) - \psi(t)] \, d\left(\frac{e^{-i\lambda t}}{-i\lambda}\right)$$

$$= \int\limits_{t_0}^{\infty} [\varphi'(t) - \psi'(t)] \, e^{-i\lambda t} \, dt$$

$$= \int\limits_{t_0}^{\infty} [(\varphi'(t) - \psi'(t))] \, e^{-Mt} \, e^{-i\lambda t + Mt} \, dt \; .$$

It is easy to see that for $M > \gamma$, the function $\left| [\varphi'(t) - \psi'(t)] \, e^{-Mt} \right|$ is integrable on the whole of t-axis. Therefore, it follows from the previous equality that for $p \equiv - \operatorname{Im} \lambda \geq M > \gamma$,

$$|\lambda[\widetilde{\varphi}(\lambda) - \widetilde{\psi}(\lambda)]| \leq e^{-t_0(p-M)} \int\limits_{t_0}^{\infty} \left| [\varphi'(t) - \psi'(t)] \, e^{-Mt} \right| dt = \mathrm{const} \, e^{-t_0 p}$$

Hence for p large enough and x bounded above

$$\left| 1 - \exp\left\{ i\lambda[\widetilde{\varphi}(\lambda) - \widetilde{\psi}(\lambda)] \frac{x}{2c} \right\} \right| \leq \mathrm{const} \, e^{-t_0 p} \qquad (3.1.16)$$

Finally, it follows from (3.1.5) that

$$\left| \exp\left[-i\lambda\widetilde{\varphi}(\lambda) \frac{x}{2c} \right] \right| \leq 1; \quad p > 0. \qquad (3.1.17)$$

Now, (3.1.15, 3.1.16, and 3.1.17) yield

$$\left| \widetilde{\varepsilon}(\lambda, x) - \frac{1}{i\lambda} \exp\left\{ -i\lambda\left[1 + \frac{\widetilde{\psi}(\lambda)}{2} \right] \frac{x}{c} \right\} \right| \leq \mathrm{const} \, \exp\left[-\left(\frac{x}{c} + t_0 \right) p \right] \qquad (3.1.18)$$

for p large enough and x bounded above. Hence by Theorem 1.3.1, one obtains the required equality as

$$\varepsilon(t, x) = F^{-1}_{\lambda \to t - (x/c)} \frac{1}{i\lambda} \exp\left\{ -i\lambda\left[1 + \frac{\widetilde{\psi}(\lambda)}{2} \right] \frac{x}{c} \right\}, \quad \text{for } t < \frac{x}{c} + t_0.$$

(b) As we know

$$\text{Im}\,[\lambda\widetilde{\varphi}(\lambda)] \leq 0, \quad \text{for } p > 0$$

(See Lemma 1.5.4). Therefore

$$\left|e^{-i\lambda\widetilde{\varphi}_1(\lambda)}\right| = \left|e^{-i\lambda\widetilde{\varphi}(\lambda)}\,e^{-i\lambda v F_{t\to\lambda}\ln t_+}\right| \leq \text{const}\,(1+|\lambda|)^v; \quad p > 0$$

whence, by Theorem 1.3.1, (3.1.12) follows.

The proof of (3.1.14) is similar to the proof of the assertion (a) of the lemma. The lemma is proved.

3.2   Two Classical Tauberian Theorems

In this section, we give formulations of two classical results which will be employed later on.

Definition   A function $g(p)$, $0 < p < \infty$, is said to be completely monotone, if it is infinitely differentiable and

$$(-1)^k \frac{d^k g(p)}{dp^k} \geq 0; \quad k = 0, 1, \ldots \tag{3.2.1}$$

Lemma 3.2.1 [2]   Let $f(p)$, $0 < p < \infty$, be a real function with a completely monotone derivative. Then, $e^{-f(p)}$ is a completely monotone function.

Proof   In accordance with the complete monotonicity of $f'(p)$, we have

$$(-1)^{n-1} f^{(n)}(p) \geq 0, \quad n = 1, 2, \ldots \tag{3.2.2}$$

for $p > 0$. Let us check that

$$(-1)^k \frac{d^k}{dp^k}\,e^{-f(p)} \geq 0, \quad k = 0, 1, \ldots \tag{3.2.3}$$

for $p > 0$. For $k = 0$, (3.2.3) is obvious. For $k = 1$, we have

$$-\frac{d}{dp}\,e^{-f(p)} = e^{-f(p)} f'(p) \geq 0$$

by virtue of (3.2.2) for $n = 1$. For $k = 2$, one obtains

$$(-1)^2 \frac{d^2}{dp^2} e^{-f(p)} = e^{-f(p)} \left(\left\{ \frac{d}{dp} [-f(p)] \right\}^2 + \frac{d^2}{dp^2} [-f(p)] \right) \geq 0$$

by virtue of (3.2.2) for $n = 2$. For $k = 3$

$$(-1)^3 \frac{d^3}{dp^3} e^{-f(p)}$$

$$= -e^{-f(p)} \left(\left\{ \frac{d}{dp} [-f(p)] \right\}^3 \right.$$

$$+ 3 \left\{ \frac{d}{dp} [-f(p)] \right\} \frac{d^2}{dp^2} [-f(p)]$$

$$\left. + \frac{d^3}{dp^3} [-f(p)] \right) \geq 0$$

by virtue of (3.2.2) for $n = 1, 2, 3$. Proceeding with differentiation, one obtains the required result.

Theorem 3.2.1 (S. Bernstein; See also [2, 3]) A function $g(p), 0 < p < \infty$, can be represented as the Laplace–Stieltjes integral

$$g(p) = \int_0^\infty e^{-pt} dG(t) \qquad (3.2.4)$$

where $G(t)$ is a non-decreasing function, which is supposed to be equal to zero for $t < 0$, if and only if $g(p)$ is completely monotone.

Definition (See [4]) A function $l(p)$ is said to be slowly varying as $p \to \infty$ if it is real valued, positive, and measurable on $[p_0, \infty]$, $p_0 > 0$ and if for each $a > 1$

$$\lim_{p \to \infty} \frac{l(ap)}{l(p)} = 1 \qquad (3.2.5)$$

Theorem 3.2.2 (J. Karamata; See also [2, 4]) Let $G(t)$ be a monotone non-decreasing function equal to zero for $t < 0$ and such that the integral

$$g(t) = \int_{0-}^\infty e^{-pt} dG(t)$$

is finite for all $p > 0$. Then

(a) If $g(p) = p^{-\rho} l(1/p)$ as $p \to +0$, then

$$G(t) \sim \frac{t^\rho l(t)}{\Gamma(\rho + 1)}, \quad \text{as } t \to +\infty; \tag{3.2.6}$$

(b) if $g(p) = p^{-\rho} l(p)$, as $p \to \infty$, then

$$G(t) \sim \frac{t^\rho l\left(\frac{1}{t}\right)}{\Gamma(\rho + 1)}, \text{ as } t \to +0 \tag{3.2.7}$$

Here Γ is the Gamma function. $l(p)$ slowly varies as p tends to infinity; $\rho > 0$.

3.3   The Continuity and Monotonicity Theorem

In this section, we prove a general result which is independent of the character of singularity of the kernel $\varphi(t)$.

Lemma 3.3.1 There exists a non-decreasing function $U(t)$ such that

$$e^{-\rho\overline{\varphi}(p)} = \int\limits_{0-}^{\infty} e^{-pt}\, dU(t), p > 0 \tag{3.3.1}$$

$U(t)$ is supposed to be equal to zero for $t < 0$.

Proof First of all, note that $p\overline{\varphi}(p)$ is non-negative, since $\varphi(t) \geq 0$. Furthermore, let us demonstrate that

$$\frac{d}{dp}[p\overline{\varphi}(p)] = -\int\limits_{0}^{\infty} t\varphi'(t)\, e^{-pt}\, dt. \tag{3.3.2}$$

Usually this formula is proved under the assumption of the function $\varphi(t)$ being bounded. To prove (3.3.2) for $\varphi(t)$ satisfying the mentioned conditions (a)–(e) from Sect. 3.1.1, let us consider the equality

$$\frac{d}{dp}\left[p\int_{\delta}^{\infty}\varphi(t)e^{-pt}dt\right] = -\frac{d}{dp}\int_{\delta}^{\infty}\varphi(t)\,de^{-pt}$$

$$= -\frac{d}{dp}[\varphi(t)e^{-pt}]\Big|_{t=\delta}^{t=\infty} + \frac{d}{dp}\int_{\delta}^{\infty}\varphi'(t)e^{-pt}dt$$

$$= -\delta\varphi(\delta)e^{-p\delta} - \int_{\delta}^{\infty}t\varphi'(t)e^{-pt}dt; \quad \delta > 0$$

Notice that there must exist a sequence $\delta_n \to +0$ such that $\delta_n\varphi(\delta_n) \to 0$, since otherwise $\varphi(t)$ would obviously have a non-integrable singularity at $t = 0$. Passing to the limit in the last equality, one easily gets the desired result. Note that we have also shown that the function $t\varphi'(t)$ is locally integrable on $[0, \infty)$.

Now, let us demonstrate that for $p > 0$, the following sequence of inequalities

$$(-1)^{k-1}\frac{d^k}{dp^k}[p\overline{\varphi}(p)] \geq 0; k = 1, 2, \ldots, \tag{3.3.3}$$

holds.

Really, (3.3.2) yields (3.3.3) for $k = 1$, as in accordance with our assumptions, $\varphi'(t) \leq 0$ for $t > 0$. Now, it is clear that

$$\frac{d^k}{dp^k}[p\overline{\varphi}(p)] = -\frac{d^{k-1}}{dp^{k-1}}\int_0^{\infty}t\varphi'(t)e^{-pt}dt = \cdots = (-1)^k\int_0^{\infty}t^k\varphi'(t)e^{-pt}dt$$

whence (3.3.3) follows.

Thus we have proved the function $[p\overline{\varphi}(p)]'$ being completely monotone. Hence $e^{-p\overline{\varphi}(p)}$ is also a completely monotone function (see Lemma 3.2.1). Whence, in accordance with Theorem 3.2.1, the stated result follows.

Lemma 3.3.2

(a) The function $U(t)$ determined in (3.3.1) can be represented as

$$U(t) = L_{p\to t}^{-1}\frac{e^{-p\overline{\varphi}(p)}}{p} \tag{3.3.4}$$

(b) Moreover $U(t) \to 0$ as $t \to +0$, and $U(t) \to 1$ as $t \to +\infty$.

Proof The assertion (a) of the lemma immediately follows from (3.3.1) after integration by parts. The assertion (b) follows from (3.3.4) by virtue of well-known limiting theorems for the Laplace transform.

Now, our purpose is to demonstrate the continuity of the function (3.3.4). To prove this fact, we have to establish two following preparatory lemmas which employ property (c) of $\varphi(t)$.

Lemma 3.3.3 Suppose a locally integrable function $\psi(t)$ equals zero both for $t < 0$ and for $t > \text{const} > 0$, is n times differentiable for $t > 0$, and for small $t \geq 0$ satisfies the condition

$$\left|\frac{d^n \psi(t)}{dt^n}\right| \leq \text{const}\, t^{-n-\delta}; \quad 0 < \delta < 1 \tag{3.3.5}$$

Then for $p = \text{const} > 0$, the entire function $\widetilde{\psi}(\lambda), \lambda = \mu - ip$, satisfies the estimates

$$\left|\frac{d^j \widetilde{\psi}(\lambda)}{d\lambda^j}\right| \leq \text{const}\,(1+|\mu|)^{-j-1+\delta+\gamma}; \quad j = 0, 1, \ldots, n \tag{3.3.6}$$

where $\gamma > 0$ can be chosen however small.

Note From the condition (3.3.5), it evidently follows that for small $t > 0$, the sequence of inequalities

$$\left|\frac{d^j \psi(t)}{dt^j}\right| \leq \text{const}\, t^{-j-\delta}; \quad j = 0, 1, \ldots, n-1, \tag{3.3.7}$$

holds.

Proof of the Lemma First of all, let us establish (3.3.6) for $j = 0$. Taking $\gamma_1 > 0$ however small and making use of the Parseval's theorem, we have

$$\widetilde{\psi}(\lambda) = \int_0^\infty t^{-\delta-\gamma_1}\, t^{\delta+\gamma_1}\, \psi(t)\, e^{-i\lambda t}\, dt$$

$$= \frac{1}{2\pi}\left(\int_0^\infty t^{-\delta-\gamma_1}\, e^{-i\mu t - pt}\, dt\right) \overset{*}{\mu} \left[\int_0^\infty t^{\delta+\gamma_1}\, \psi(t)\, e^{-i\mu t}\, dt\right] \tag{3.3.8}$$

where $\overset{*}{\mu}$ denotes convolution with respect to μ.

Let us evaluate the first integral on the right-hand side of (3.3.8). We have

$$\int_0^\infty t^{-\delta-\gamma_1}\, e^{-i\mu t - pt}\, dt = \Gamma(1 - \delta - \gamma_1)\,(p + i\mu)^{\delta+\gamma_1-1}$$

where Γ is the Gamma function. Whence

$$\left| \int\limits_0^\infty t^{-\delta-\gamma_1} e^{-i\mu t - pt} \, dt \right| \le \text{const} \, |\mu|^{\delta+\gamma_1-1} \tag{3.3.9}$$

Furthermore, let us evaluate the second integral on the right-hand side of (3.3.8). Note that by virtue of (3.3.7) with $j = 0$

$$t^{\delta+\gamma_1} \psi(t) \to 0, \quad \text{as } t \to +0.$$

Therefore, integration by parts yields

$$\int\limits_0^\infty t^{\delta+\gamma_1} \psi(t) \, e^{-i\mu t} dt = \frac{1}{i\mu} \int\limits_0^\infty \frac{d}{dt} \left[t^{\delta+\gamma_1} \psi(t) \right] dt. \tag{3.3.10}$$

On the one hand, the module of the left-hand side of (3.3.10) is obviously bounded above by a constant independent of μ. On the other hand, the module of the right-hand side of (3.3.10) is bounded above by const/ $|\mu|$, where const is again independent of μ. Thus we arrive at the inequality

$$\left| \int\limits_0^\infty t^{\delta+\gamma_1} \psi(t) \, e^{-i\mu t} \, dt \right| \le \frac{\text{const}}{1 + |\mu|}. \tag{3.3.11}$$

Now, (3.3.8, 3.3.9, 3.3.11) yield

$$
\begin{aligned}
|\widetilde{\psi}(\lambda)| &\le \text{const} \int\limits_{-\infty}^\infty |\xi|^{\delta+\gamma_1-1} \frac{d\xi}{1 + |\mu - \xi|} \\
&\le \text{const} \int\limits_{-\infty}^\infty |\xi|^{\delta+\gamma_1-1} |\mu - \xi|^{\gamma_2-1} \, d\xi \\
&\le \text{const} |\mu|^{\delta+\gamma_1+\gamma_2-1}
\end{aligned}
\tag{3.3.12}
$$

Here $\gamma_2 > 0$ is however small. But $\widetilde{\psi}(\lambda)$ is evidently a bounded function of μ for $p = \text{const} > 0$. Hence, it follows from (2.3.12) that for $p = \text{const} > 0$.

$$|\widetilde{\psi}(\lambda)| \le \text{const} \, (1 + |\mu|)^{\delta+\gamma_1+\gamma_2-1} \tag{3.3.13}$$

Redenoting $\gamma_1 + \gamma_2$ by γ, we arrive at the desired result. Now, let us establish (3.3.6) for $j \ge 1$. We have

$$\frac{\mathrm{d}^j}{\mathrm{d}\lambda^j}\widetilde{\psi}(\lambda) = \int_0^\infty (-it)^j \psi(t) e^{-i\lambda t} \mathrm{d}t \tag{3.3.14}$$

But by virtue of (3.3.7)

$$\frac{\mathrm{d}^k}{\mathrm{d}t^k}\left[t^j\psi(t)\right] \to 0, \quad \text{as } t \to +0; \quad k = 1, 2, \ldots, j-1$$

Hence, by integration by parts, we successively obtain from (3.3.14)

$$\frac{\mathrm{d}^j}{\mathrm{d}\lambda^j}\widetilde{\psi}(\lambda) = (-i)^j \int_0^\infty t^j \psi(t)\, \mathrm{d}\frac{e^{-i\lambda t}}{-i\lambda}$$

$$= -\frac{(-i)^j}{-i\lambda} \int_0^\infty \frac{\mathrm{d}}{\mathrm{d}t}\left[t^j\psi(t)\right] e^{-i\lambda t}\, \mathrm{d}t \tag{3.3.15}$$

$$=$$
$$\vdots$$

$$= (-1)^j \frac{(-i)^j}{(-i\lambda)^j} \int_0^\infty \frac{\mathrm{d}^j}{\mathrm{d}t^j}\left[t^j\psi(t)\right] e^{-i\lambda t}\, \mathrm{d}t$$

As before, let us choose $\gamma_1 > 0$, however small. Then for $p = \text{const} > 0$, we can rewrite the integral on the right-hand side of (3.3.15) as follows

$$\int_0^\infty \frac{\mathrm{d}^j}{\mathrm{d}t^j}\left[t^j\psi(t)\right] e^{-i\lambda t}\, \mathrm{d}t = \int_0^\infty t^{-\delta-\gamma_1} t^{\delta+\gamma_1} \frac{\mathrm{d}^j}{\mathrm{d}t^j}\left[t^j\psi(t)\right] e^{i\lambda t}\, \mathrm{d}t$$

$$= \frac{1}{2\pi}\left(\int_0^\infty t^{-\delta-\gamma_1} e^{-i\mu t - pt}\, \mathrm{d}t\right) \overset{*}{\mu}\left(\int_0^\infty t^{\delta+\gamma_1} \frac{\mathrm{d}^j}{\mathrm{d}t^j}\left[t^j\psi(t)\right] e^{-i\mu t}\, \mathrm{d}t\right) \tag{3.3.16}$$

Note that by virtue (3.3.7)

$$t^{\delta+\gamma_1} \frac{\mathrm{d}^j}{\mathrm{d}t^j}\left[t^j\psi(t)\right] \to 0, \quad \text{as } t \to +0.$$

Hence one can use integration by parts in the second integral on the right-hand side of (3.3.16). Now, evaluation of (3.3.16), which is just like in the case of $j = 0$ and formula (3.3.15), give the required result.

Lemma 3.3.4 For $p > 0$

$$e^{-p\overline{\varphi}(p)} = \int_0^\infty u(t)\, e^{-pt}\, dt \qquad (3.3.17)$$

where $u(t)$ is a locally integrable function.

Proof By virtue of Lemma 3.1.1, one can evidently consider the function $\varphi(t)$ as having a compact support. Furthermore, by virtue of the same Lemma 3.1.1

$$F_{\lambda \to t}^{-1}\, e^{-i\lambda\widetilde{\varphi}(\lambda)} = 0,\, \text{for } t < 0. \qquad (3.3.18)$$

Now, supposing $p > 0$, consider the expression

$$F_{t \to \lambda}\left[t^n\, F_{\lambda \to t}^{-1}\, e^{-i\lambda\widetilde{\varphi}(\lambda)} \right] = \left(i\frac{d}{d\lambda} \right)^n e^{-i\lambda\widetilde{\varphi}(\lambda)} \qquad (3.3.19)$$

As we know, by virtue of Lemma 1.5.4, which obviously holds true for the function $\varphi(t)$

$$\text{Im}\left[\lambda\widetilde{\varphi}(\lambda)\right] \leq 0, \quad \text{for } p > 0.$$

Furthermore, by virtue of the previous lemma

$$\left| \frac{d^j\widetilde{\varphi}(\lambda)}{d\lambda^j} \right| \leq \text{const}\, (1 + |\mu|)^{-j-1+\delta+\gamma}; j = 0, 1, \ldots, n$$

for $p \geq \text{const} > 0$, where $\gamma > 0$ is however small. Then one easily obtains

$$\left| \left(i\frac{d}{d\lambda} \right)^n e^{-i\lambda\widetilde{\varphi}(\lambda)} \right| \leq \text{const}\, (1 + |\mu|)^{-n(1-\delta-\gamma)} \qquad (3.3.20)$$

for $p \geq \text{const} > 0$. Since by property (c) of the function $\varphi(t)$ (see Sect. 3.1) $n (1 - \delta - \gamma) > 1/2$ for $\gamma > 0$ small enough, it follows from (3.3.20) that the function (3.3.19) turns out to be square integrable over

$$d\mu \text{ on } (-\infty, \infty) \text{ for } p = \text{const} > 0.$$

Hence the only possible singularity of the function $F_{\lambda \to t}^{-1} e^{-i\lambda \widetilde{\varphi}(\lambda)}$ may be located at $t = 0$. As is well-known from the theory of distributions, the mentioned singularity must have the form

$$\sum_{k=0}^{s} c_k \delta^{(k)}(t)$$

where $\delta(t)$ is the Delta function.

Thus in accordance with (3.3.18), one can write

$$e^{-i\lambda \widetilde{\varphi}(\lambda)} = \int_{0}^{\infty} u(t) e^{-i\lambda t} \, \mathrm{d}t + \sum_{k=0}^{s} c_k (i\lambda)^k \qquad (3.3.21)$$

where $u(t)$ is a locally integrable function such that $e^{-\gamma t} u(t)$ is Lebesgue integrable on $[0, \infty)$ for however small $\gamma > 0$. Hence by putting in (3.3.21), $\lambda = -ip, p > 0$, we obtain

$$e^{-p\overline{\varphi}(p)} = \int_{0}^{\infty} u(t) e^{-pt} \, \mathrm{d}t + \sum_{k=0}^{s} c_k p^k. \qquad (3.3.22)$$

Now, from the fact that $\varphi(t) \to \infty$ as $t \to +0$, from the property (d) of the function $\varphi(t)$, it follows that

$$p\overline{\varphi}(p) \to \infty, \quad \text{as } p \to \infty.$$

Therefore, letting in (3.3.22) $p \to \infty$, one easily obtains

$$c_k = 0; k = 0, 1, \ldots, s.$$

Hence follows the required result.

Lemma 3.3.5 The function $U(t)$ determined by (3.3.1) (or, which is the same, by (3.3.4)) is continuous for all t.

Proof Comparing formulas (3.3.4) and (3.3.17), one can easily see that

$$U(t) = \int_{0}^{t} u(\tau) \, \mathrm{d}\tau \qquad (3.3.23)$$

whence by virtue of Lemma 3.3.4, follows the assertion of the lemma.

Corollary In the representation (3.3.17)

$$u(t) \geq 0 \tag{3.3.24}$$

for almost all $t \geq 0$.

Lemma 3.3.6 Let $0 \leq a \leq b$ Then

$$L_{p \to t}^{-1} \frac{e^{-p\bar{\varphi}(p)a}}{p} \geq L_{p \to t}^{-1} \frac{e^{-p\bar{\varphi}(p)b}}{p}. \tag{3.3.25}$$

Proof At first, let $a = 0$. Then (3.3.25) easily follows from Lemmas 3.3.1 and 3.3.2. Now, let $a > 0$. Then we have

$$L_{p \to t}^{-1} \left[\frac{e^{-p\bar{\varphi}(p)a}}{p} - \frac{e^{-p\bar{\varphi}(p)b}}{p} \right]$$

$$= L_{p \to t}^{-1} e^{-p\bar{\varphi}(p)a} \left[\frac{1}{p} - \frac{e^{-p\bar{\varphi}(p)(b-a)}}{p} \right] \tag{3.3.26}$$

$$= \left[L_{p \to t}^{-1} e^{-p\tilde{\varphi}(p)a} \right] * \left\{ L_{p \to t}^{-1} \left[\frac{1}{p} - \frac{e^{-p\bar{\varphi}(p)(b-a)}}{p} \right] \right\}$$

By virtue of Corollary to Lemma 3.3.4

$$L_{p \to t}^{-1} e^{-p\bar{\varphi}(p)a} \geq 0.$$

Furthermore, it easily follows from Lemmas 3.3.1 and 3.3.2 that for $t > 0$

$$L_{p \to t}^{-1} \left[\frac{1}{p} - \frac{e^{-p\bar{\varphi}(p)(b-a)}}{p} \right] \equiv 1 - L_{p \to t}^{-1} \frac{e^{-p\bar{\varphi}(p)(b-a)}}{p} \geq 0.$$

Hence the right-hand side of (3.3.26) presents a convolution of two non-negative functions, which gives the result required.

Theorem 3.3.1 For $x > 0$, the solution of the problem (3.1.1, 3.1.2, and 3.1.3)

$$\varepsilon(t, x) = L_{p \to t - (x/c)}^{-1} \frac{1}{p} \exp\left[-p\bar{\varphi}(p) \frac{x}{2c} \right],$$

Possesses the following properties:

(a) $\varepsilon(t, x)$ is a non-decreasing in t function.
(b) $\varepsilon(t, x)$ is a non-increasing in x function.

$$\lim_{t \to (x/c)+0} \varepsilon(t,x) = 0, \tag{3.3.27}$$

$$\lim_{t \to \infty} \varepsilon(t,x) = 1; \tag{3.3.28}$$

(c) $\varepsilon(t,x)$ is continuous in t, x.

Proof The assertions (a) and (c) of the theorem directly follow from Lemmas 3.3.1 and 3.3.2. As regards the assertion (b), let us note the following:

By virtue of Lemma 3.3.6, for each $t_0 > 0, \varepsilon(\frac{x}{c} + t_0, x)$ proves to be a non-increasing in x function. Let $x_2 > x_1 > 0$, then

$$\varepsilon\left(\frac{x_1}{c} + t_0, x_1\right) \geq \varepsilon\left(\frac{x_2}{c} + t_0, x_2\right). \tag{3.3.29}$$

On the other hand, by virtue of the assertion (a) of the theorem

$$\varepsilon\left(\frac{x_1}{c} + t_0, x_1\right) \leq \varepsilon\left(\frac{x_2}{c} + t_0, x_1\right). \tag{3.3.30}$$

Hence it turns out from (3.3.29, 3.3.30) that

$$\varepsilon\left(\frac{x_2}{c} + t_0, x_1\right) \geq \varepsilon\left(\frac{x_2}{c} + t_0, x_2\right).$$

Redenoting $(x_2/c) + t_0$ by t gives the result required.

Finally, the assertion (d) of the theorem easily follows from Lemma 3.3.5 and assertions (a) and (b).

3.4   The Approximation Theorem

Theorem 3.4.1 Let functions $\varphi_j(t)$, $j = 1, 2, \ldots$ be smooth, non-increasing, and concave for $t > 0$, integrable $[0, \infty)$ and let $\varphi_j(t) = 0$ for $t < 0$. Furthermore, suppose

$$\int_0^\infty \left|\varphi(t) - \varphi_j(t)\right| dt \to 0, \quad \text{as } j \to \infty. \tag{3.4.1}$$

Then for each fixed x

$$\varepsilon(t,x) - F_{\lambda \to t-(x/c)}^{-1} \left\{ \frac{1}{i\lambda} \exp\left[-i\lambda \tilde{\varphi}_j(\lambda) \frac{x}{2c} \right] \right\} \to 0, \quad \text{as } j \to \infty \qquad (3.4.2)$$

uniformly with respect to t.

Proof Let us denote

$$\varepsilon_j(t,x) \equiv F_{\lambda \to t-(x/c)}^{-1} \left\{ \frac{1}{i\lambda} \exp\left[-i\lambda \tilde{\varphi}_j(\lambda) \frac{x}{2c} \right] \right\}$$

First of all, let us demonstrate that for however small $\delta > 0$

$$\int_{-\infty}^{\infty} e^{-2\delta t} |\varepsilon(t,x) - \varepsilon_j(t,x)|^2 \, dt \to 0 \ \text{as } j \to \infty. \qquad (3.4.3)$$

We have

$$\tilde{\varepsilon}(\lambda,x) - \tilde{\varepsilon}_j(\lambda,x) = \frac{1}{i\lambda} \exp\left\{ -i\lambda \left[1 + \tilde{\varphi}\left(\frac{\lambda}{2}\right) \right] \frac{x}{c} \right\}$$
$$\times \left(1 - \exp\left\{ i\lambda [\tilde{\varphi}(\lambda) - \tilde{\varphi}_j(\lambda)] \frac{x}{2c} \right\} \right) \qquad (3.4.4)$$

By virtue of (3.4.1)

$$\tilde{\varphi}(\lambda) - \tilde{\varphi}_j(\lambda) \to 0 \ \text{as } j \to \infty \qquad (3.4.5)$$

uniformly in the half-plane $p \geq \delta > 0$. Hence for $p = \delta$, the difference (3.4.4) goes to zero uniformly on each segment $-M \leq \mu \leq M$, as $j \to \infty$.

Furthermore, as we know

$$\text{Im}\, [\lambda \tilde{\varphi}(\lambda)] \leq 0, \quad \text{for } p > 0$$

(see the proof of Lemma 1.5.4), whence

$$|\tilde{\varepsilon}(\lambda,x)| \leq \frac{1}{|\lambda|}, \text{for } p > 0 \qquad (3.4.6)$$

Similarly, one can prove that

$$\text{Im}\big(\lambda \tilde{\varphi}_j(\lambda)\big) \leq 0 \,\text{for } p > 0$$

whence

$$|\widetilde{\varepsilon}_j(\lambda, x)| \le \frac{1}{|\lambda|}, \quad \text{for } p > 0 \tag{3.4.7}$$

Now, from (3.4.4, 3.4.5, 3.4.6, and 3.4.7), one easily gets that for $p = \delta > 0$,

$$\int_{-\infty}^{\infty} |\widetilde{\varepsilon}(\lambda, x) - \widetilde{\varepsilon}_j(\lambda, x)|^2 \, d\mu \to 0, \quad \text{as } j \to \infty$$

whence by Plancherel's theorem, (3.4.3) follows.

Furthermore, as we know from Theorem 3.3.1, for $x > 0$, $\varepsilon(t, x)$ is a continuous, non-decreasing in t, bounded above function equal to zero for $t < x/c$. In the same manner, one can prove that for $x > 0$ all $\varepsilon_j(t, x)$ are non-decreasing in t bounded above functions equal to zero for $t < x/c$. Since the values $\varphi_j(+0)$ may be finite, the functions $\varepsilon_j(t, x)$ may suffer jumps across the wave front $t = x/c$.

Now, from above-mentioned, follows the required result for $x > 0$. Finally, in case of $x = 0$, the assertion of the theorem is evident.

Note To evaluate the difference $\varepsilon(t, x) - \varepsilon_j(t, x)$, one can make use of the well-known Esseen's inequality (see [5]).

Note One can easily see that for $\varphi(t)$ tending to a finite limit as $t \to +0$ Theorem 3.4.1 is valid only under the additional assumption of

$$\varphi_j(+0) \to \varphi(+0), \quad \text{as } j \to \infty.$$

3.5 Memory Function with a Singularity, Which Is Weaker than the Logarithmic One

In Sects. 3.5 and 3.6, we have studied the wave front asymptotics of $\varepsilon(t, x)$ by purely real methods. As usual, the function $\varphi(t)$ is supposed satisfying conditions (a)–(e) in Sect 3.1.

Lemma 3.5.1 (Abelian) Let

$$\varphi'(t) = o\left(\frac{1}{t}\right), \quad \text{as } t \to +0. \tag{3.5.1}$$

Then

$$p\bar{\varphi}(p) - \varphi\left(\frac{1}{p}\right) \to 0, \quad \text{as } p \to +\infty \qquad (3.5.2)$$

Note From the condition (3.5.1), it evidently follows that

$$\varphi(t) = o\left(\ln\frac{1}{t}\right), \text{ as } t \to +0.$$

Proof of the Lemma We have

$$p\bar{\varphi}(p) - \varphi\left(\frac{1}{p}\right) = p\int_0^\infty \varphi(t)e^{-pt}\,dt - \varphi\left(\frac{1}{p}\right)$$

$$= \int_0^\infty \varphi\left(\frac{t}{p}\right)e^{-t}\,dt - \varphi\left(\frac{1}{p}\right)$$

$$= \int_0^1 \left[\varphi\left(\frac{t}{p}\right) - \varphi\left(\frac{1}{p}\right)\right]e^{-t}\,dt + \int_1^\infty \left[\varphi\left(\frac{t}{p}\right) - \varphi\left(\frac{1}{p}\right)\right]e^{-t}\,dt$$

$$(3.5.3)$$

Let

$$I_1 = \int_0^1 \left[\varphi\left(\frac{t}{p}\right) - \varphi\left(\frac{1}{p}\right)\right]e^{-t}\,dt,$$

$$I_2 = \int_1^\infty \left[\varphi\left(\frac{t}{p}\right) - \varphi\left(\frac{1}{p}\right)\right]e^{-t}\,dt.$$

Let us evaluate the integral I_1. For $0 < t \le 1$, we have

$$\varphi\left(\frac{t}{p}\right) - \varphi\left(\frac{1}{p}\right) \ge 0$$

since $\varphi(t)$ is non-increasing for $t > 0$. Therefore

$$0 \le I_1 \le \int_0^1 \left[\varphi\left(\frac{t}{p}\right) - \varphi\left(\frac{1}{p}\right) \right] dt$$

$$= p \int_0^{1/p} \varphi(\tau) d\tau - \varphi\left(\frac{1}{p}\right)$$

$$= p\tau\varphi(\tau)\big|_{\tau=0}^{\tau=\frac{1}{p}} - p \int_0^{1/p} \tau\varphi'(\tau) d\tau - \varphi\left(\frac{1}{p}\right)$$

$$= -p \int_0^{1/p} \tau\varphi'(\tau) d\tau$$

From (3.5.1), it is clear that

$$-p \int_0^{1/p} \tau\varphi'(\tau) d\tau \to 0, \quad \text{as } p \to +\infty$$

Hence I_1 also goes to zero as $p \to +0$.

Now, let us evaluate the integral I_2. For $t \ge 1$, we have

$$0 \ge \varphi\left(\frac{t}{p}\right) - \varphi\left(\frac{1}{p}\right) = \int_{1/p}^{t/p} \varphi'(\tau) d\tau \ge \frac{t}{p}\varphi'\left(\frac{1}{p}\right) \tag{3.5.4}$$

as $\varphi(t)$ is non-increasing and concave for $t > 0$. From (3.5.4), it follows that

$$0 \ge I_2 \ge \frac{1}{p}\varphi'\left(\frac{1}{p}\right) \int_1^\infty te^{-t} dt$$

whence, by virtue of (3.5.1), it is clear that $I_2 \to 0$ as $p \to +\infty$.

Thus we have demonstrated that the difference (3.5.3) goes to zero as $p \to +\infty$. The lemma is proved.

In Sect. 3.6, we shall make use of the following variant of the previous lemma.

Lemma 3.5.1′ Let $\varphi_1(t)$, $t \in (0, +\infty)$, be a function whose Laplace–transform makes sense for $p > 0$ and suppose

$$\varphi_1\left(\frac{t}{p}\right) - \varphi_1\left(\frac{1}{p}\right) \to 0, \quad \text{as } p \to +\infty \qquad (3.5.5)$$

uniformly with respect to t on each segment of the sort $0 < a \leq t \leq b$;

(a) For large $p > 0$

$$\left|\varphi_1\left(\frac{t}{p}\right) - \varphi_1\left(\frac{1}{p}\right)\right| \leq \psi(t) \qquad (3.5.6)$$

where $\psi(t) \geq 0$ is some function whose Laplace transform makes sense for $p > 0$. Then

$$p\overline{\varphi}_1(p) - \varphi_1\left(\frac{1}{p}\right) \to 0, \quad \text{as } p \to +\infty. \qquad (3.5.7)$$

Lemma 3.5.2 (Tauberian) Let $\varphi(t)$ satisfy the condition of Lemma 3.5.1. Then the function $e^{-p\overline{\varphi}(p)}$ slowly varies as $p \to \infty$ and

$$L_{p\to t}^{-1} \frac{e^{-p\overline{\varphi}(p)}}{p} \sim e^{-\varphi(t)}, \quad \text{as } t \to +0. \qquad (3.5.8)$$

Proof For $t > 0$, we have

$$\frac{1}{e^{-p\overline{\varphi}(p)}} \exp\left[-\frac{p}{t}\overline{\varphi}\left(\frac{p}{t}\right)\right] = \exp\left[-\frac{p}{t}\overline{\varphi}\left(\frac{p}{t}\right) + p\overline{\varphi}(p)\right]$$
$$= \exp\left\{-\left[\varphi\left(\frac{t}{p}\right) - \varphi\left(\frac{1}{p}\right)\right] + o(1)\right\} \qquad (3.5.9)$$

by virtue of Lemma 3.5.1. Let us demonstrate that

$$\varphi\left(\frac{t}{p}\right) - \varphi\left(\frac{1}{p}\right) \to 0, \quad \text{as } p \to +\infty. \qquad (3.5.10)$$

In fact, in case of $0 < t \leq 1$, we have

$$0 \leq \varphi\left(\frac{t}{p}\right) - \varphi\left(\frac{1}{p}\right) = -\int\limits_{t/p}^{1/p} \varphi'(\tau)\,d\tau$$

$$\leq -\varphi'\left(\frac{t}{p}\right)\frac{1-t}{p}$$

$$\leq -\varphi'\left(\frac{t}{p}\right)\frac{1}{p} = -\frac{t}{p}\varphi'\left(\frac{t}{p}\right)\frac{1}{t} \to 0, \quad \text{as } p \to +\infty$$

by virtue of (3.5.1), which proves (3.5.10). In case of $t > 1$, (3.5.10) follows, for example, from (3.5.4). So, it follows from (3.5.10) that the ratio on the left-hand side of (3.5.9) tends to 1, as $p \to +\infty$, that is $e^{-p\overline{\varphi}(p)}$ slowly varies as $p \to +\infty$.

Furthermore, as we know from the Note to Lemmas 3.3.1 and 3.3.2

$$e^{-p\overline{\varphi}(p)} = \int\limits_0^\infty e^{-pt}\,dU(t) \tag{3.5.11}$$

where $U(t)$ is a monotone non-decreasing function, whence by Karamata's theorem, see formula (3.2.5)

$$U(t) \sim \exp\left[-\frac{1}{t}\overline{\varphi}\left(\frac{1}{t}\right)\right], \text{as } t \to +0..$$

But by virtue of Lemma 3.5.1

$$\frac{1}{t}\overline{\varphi}\left(\frac{1}{t}\right) - \varphi(t) \to 0, \quad \text{as } t \to +0.$$

Therefore

$$U(t) \sim e^{-\varphi(t)}, \quad \text{as } t \to +0. \tag{3.5.12a}$$

But, as we know, (3.5.11) can be rewritten as

$$U(t) = L^{-1}_{p \to t} \frac{e^{-p\overline{\varphi}(p)}}{p}$$

whence follows the required result.

Theorem 3.5.1 Let

$$\varphi'(t) = o\left(\frac{1}{t}\right), \text{ as } t \to +0$$

Then

$$\varepsilon(t,x) \sim \exp\left[-\varphi\left(t - \frac{x}{c}\right)\frac{x}{2c}\right], \quad \text{as } t \to \frac{x}{c} + 0. \tag{3.5.13}$$

Proof Formula (3.5.13) immediately follows from Lemma 3.5.2.

Now let us pass to the evaluation of $\varepsilon(t,x)$ by using the method of Tauberian inequalities.

Lemma 3.5.3 Let (3.5.1) be valid, and suppose $-t\varphi'(t)$ increases for $0 < t < t_0$, where $t_0 > 0$. Then

$$L_{p \to t}^{-1} \frac{e^{-p\bar{\varphi}(p)}}{p} \geq e^{-\varphi(t)}, \quad \text{for } 0 \leq t < t_0. \tag{3.5.14}$$

Proof As we know from Sect. 3.3

$$u(t) \equiv L_{p \to t}^{-1} e^{-p\bar{\varphi}(p)} \geq 0, \quad \text{for } t \geq 0 \tag{3.5.15}$$

See Eq. (3.3.24). Furthermore, by means of the Laplace transform, one can easily verify that for $t \geq 0$ the function $u(t)$ satisfies the following integral Volterra type equation

$$tu(t) = -\int_0^t \tau\varphi'(\tau)\,u(t-\tau)\,d\tau. \tag{3.5.16}$$

Note that by virtue of property (b) of Sect. 3.1, $\varphi'(t) \leq 0$. Therefore, taking into account (3.3.15) and the increase and non-negativeness of $-t\varphi'(t)$, one easily obtains from (3.5.16) that

$$tu(t) \leq -t\varphi'(t)\int_0^t u(\tau)\,d\tau, \quad \text{for } 0 < t < t_0$$

or, which is the same

$$\frac{d}{dt}[\varphi(t) + \ln U(t)] \leq 0, \quad \text{for } 0 < t < t_0 \tag{3.5.17}$$

where

$$U(t) = \int_0^t u(\tau)d\tau = L_{p \to t}^{-1} \frac{e^{-p\bar{\varphi}(p)}}{p}$$

Hence, by integration over dt along the interval (γ, t), where $0 < \gamma < t < t_0$, one easily obtains

$$\int_\gamma^t \frac{d}{dt}[\varphi(t) + \ln U(t)]dt = \varphi(t) + \ln U(t) - \varphi(\gamma) - \ln U(\gamma) \leq 0$$

whence

$$U(t) \leq e^{-\varphi(t)} U(\gamma) e^{\varphi(\gamma)}. \tag{3.5.18}$$

However, by virtue of Lemma 3.5.2

$$U(\gamma) e^{\varphi(\gamma)} \to 1, \quad \text{as } \gamma \to +0.$$

Therefore, letting $\gamma \to +0$ in (3.5.18), we obtain the desired estimate

$$U(t) \leq e^{-\varphi(t)}, \quad 0 \leq t < t_0.$$

The Lemma is proved. Thus we arrive at the following result.

Theorem 3.5.2 Let (3.5.1) be valid and suppose $-t\varphi'(t)$ increases for $0 < t < t_0$, where $t_0 > 0$. Then

$$0 \leq \varepsilon(t, x) \leq \exp\left[-\varphi\left(t - \frac{x}{c}\right)\frac{x}{2c}\right], \quad \text{for } \frac{x}{c} \leq t < \frac{x}{c} + t_0. \tag{3.5.19}$$

3.6   Memory Function with the Logarithmic Singularity

Now, let us pass to the case where the creep kernel has a singularity (as $t \to +0$) of the logarithmic type. As we shall see below, the asymptotic behaviour of the solution of the problem (3.1.1, 3.1.2, and 3.1.3)

$$\varepsilon(t,x) = L_{p \to t - \frac{x}{c}}^{-1} \frac{e^{-p\bar\varphi(p)\frac{x}{2c}}}{p},\tag{3.6.1}$$

has a somewhat unexpected character.

Theorem 3.6.1 Let

$$\varphi(t) = k \ln \frac{1}{t}, \quad \text{for } 0 < t < t_0\tag{3.6.2}$$

where $t_0 > 0$, $k > 0$. Then

$$\varepsilon(t,x) = \frac{e^{-kCx/(2c)}}{\Gamma\left(\frac{kx}{2c} + 1\right)} \left(t - \frac{x}{c}\right)^{kx/(2c)}, \quad \text{for } \frac{x}{c} \le t \le \frac{x}{c} + t_0\tag{3.6.3}$$

where $C = 0.57...$ is the Euler's constant, Γ the Gamma function..

Proof By virtue of Lemma 3.1.1, when calculating $\varepsilon(t,x)$ for $t \le (x/c) + t_0$, one can consider

$$\varphi(t) = k \ln \frac{1}{t}\tag{3.6.4}$$

for all $t > 0$. Then

$$\bar\varphi(p) = k \frac{C + \ln p}{p}.\tag{3.6.5}$$

After inserting (3.6.5) into the right-hand side of (3.6.1), we have for $t < (x/c) + t_0$,

$$\varepsilon(t,x) = L_{p \to t - (x/c)}^{-1} \frac{1}{p} \exp\left(-p \frac{k}{2} \frac{C + \ln p}{p} \frac{x}{c}\right)$$

whence follows the required result.

Note It follows from (3.6.3) that $\varepsilon(t,x)$ becomes more smooth in the vicinity of the wave front $t = x/c$ with the growth of x.

Now we shall extend the result of Theorem 3.6.1 to a class of memory functions, which are equivalent to $k \ln (1/t)$ but do not necessarily coincide with $k \ln (1/t)$ for small $t > 0$.

Lemma 3.6.1 (Abelian) Let for $t > 0$

$$\varphi\left(\frac{t}{p}\right) - \varphi\left(\frac{1}{p}\right) \to k \ln \frac{1}{t}, \quad \text{as } p \to +\infty \tag{3.6.6}$$

where $k > 0$, and let for large $p > 0$

$$\left|\varphi\left(\frac{t}{p}\right) - \varphi\left(\frac{1}{p}\right)\right| \le f(t) \tag{3.6.7}$$

where $f(t) \ge 0$ is a function whose Laplace transform is determined for $p > 0$. Then

$$p\overline{\varphi}(p) - \varphi\left(\frac{1}{p}\right) \to kC, \quad \text{as } p \to +\infty \tag{3.6.8}$$

Here $C = 0.57\ldots$ is the Euler's constant.

Proof Let

$$\varphi_1(t) = \varphi(t) - k \ln \frac{1}{t} \tag{3.6.9}$$

Let us demonstrate that $\varphi_1(t)$ satisfies the conditions of Lemma 3.5.1'. In fact, by virtue of (3.6.6)

$$\varphi_1\left(\frac{t}{p}\right) - \varphi_1\left(\frac{1}{p}\right) = \left[\varphi\left(\frac{t}{p}\right) - \varphi\left(\frac{1}{p}\right)\right] - k\left(\ln \frac{p}{t} - \ln p\right)$$

$$= \varphi\left(\frac{t}{p}\right) - \varphi\left(\frac{1}{p}\right) - k \ln \frac{1}{t} \to 0, \quad p \to \infty,$$

uniformly with respect to t on each segment of the sort $0 < a \le t \le b$, since the function $\varphi(t/p) - \varphi(1/p)$ is monotone in t and the function $k \ln(1/t)$ is continuous for $t > 0$. Furthermore

$$\left|\varphi_1\left(\frac{t}{p}\right) - \varphi_1\left(\frac{1}{p}\right)\right| \le \left|\varphi\left(\frac{t}{p}\right) - \varphi\left(\frac{1}{p}\right)\right| + k\left|\ln \frac{p}{t} - \ln p\right|$$

$$\le f(t) + k\left|\ln \frac{1}{t}\right|.$$

From the last two formulas, it follows that the conditions of Lemma 3.5.1' are satisfied. Therefore

$$p\bar{\varphi}_1(p) - \varphi_1\left(\frac{1}{p}\right) \to 0, \quad \text{as } p \to +\infty.$$

However by virtue of (3.6.9)

$$p\bar{\varphi}_1(p) - \varphi_1\left(\frac{1}{p}\right) = \left[p\bar{\varphi}(p) - pk\frac{C + \ln p}{p}\right] - \left[\varphi\left(\frac{1}{p}\right) - k\ln p\right]$$

$$= p\bar{\varphi}(p) - \varphi\left(\frac{1}{p}\right) - kC,$$

which proves the lemma.

The condition (3.6.6) imposed above on the function $\varphi(t)$ may seem somewhat artificial. To dispel this impression, we adduce the following modification of a proposition taken from [4].

Proposition (See [4]) Let $\varphi(t)$ be defined for $t > 0$ and suppose that for each $t > 0$, there exists

$$\lim_{p \to +\infty}\left[\varphi\left(\frac{t}{p}\right) - \varphi\left(\frac{1}{p}\right)\right] = f(t) \tag{3.6.10}$$

where $f(t)$ is a continuous function. Then $f(t) = \text{const.} \ln t$.

Proof Consider the expression

$$\left[\varphi\left(\frac{t_1}{p}\right) - \varphi\left(\frac{1}{p}\right)\right] - \left[\varphi\left(\frac{t_2}{p}\right) - \varphi\left(\frac{1}{p}\right)\right]. \tag{3.6.11}$$

On the one hand, from (3.6.10), it is clear that (3.6.11) tends to $f(t_1) - f(t_2)$ as $p \to +\infty$. On the other hand, by opening the brackets in (3.6.11) and letting $q = p/t_2$, one obtains

$$\varphi\left(\frac{t_1}{p}\right) - \varphi\left(\frac{t_2}{p}\right) = \varphi\left(\frac{t_1/t_2}{q}\right) - \varphi\left(\frac{1}{q}\right) \to f\left(\frac{t_1}{t_2}\right), \quad \text{as } q \to +\infty$$

whence

$$f\left(\frac{t_1}{t_2}\right) = f(t_1) - f(t_2),$$

which gives the required result.

Lemma 3.6.2 (Tauberian): Suppose $\varphi(t)$ satisfies the conditions of Lemma 3.6.1. Then the function

$$l(p) \equiv p^k e^{-p\overline{\varphi}(p)}$$

slowly varies as $p \to +\infty$ and

$$L_{p \to t}^{-1} \frac{l(p)}{p^{k+1}} \equiv L_{p \to t}^{-1} \frac{e^{-p\overline{\varphi}(p)}}{p} \sim \frac{e^{-kC}}{\Gamma(k+1)} e^{-\varphi(t)}, \quad t \to +0. \tag{3.6.12}$$

Here $C = 0.57...$ is the Euler's constant, the constant $k > 0$ is defined in (3.6.6).

Proof We have

$$e^{kC}l(p) \equiv e^{kC} p^k e^{-p\overline{\varphi}(p)} = \exp\left\{ -p\left[\overline{\varphi}(p) - k\frac{C + \ln p}{p} \right] \right\} = e^{-p\overline{\varphi}_1(p)}$$

where

$$\varphi_1(t) = \varphi(t) - k\ln\frac{1}{t}.$$

As was noted in the proof of Lemma 3.6.1, the function $\varphi_1(t)$ satisfies the conditions of Lemma 3.5.1'. Therefore by virtue of Lemma 3.5.1'

$$\frac{l\left(\frac{p}{t}\right)}{l(p)} = \frac{e^{kC} l\left(\frac{p}{t}\right)}{e^{kC} l(p)}$$

$$= \frac{1}{e^{-p\overline{\varphi}_1(p)}} \exp\left[\frac{-p}{t}\overline{\varphi}_1\left(\frac{p}{t}\right) \right]$$

$$= \exp\left\{ -\left[\varphi_1\left(\frac{t}{p}\right) - \varphi_1\left(\frac{1}{p}\right) \right] + o(1) \right\},$$

as $p \to +\infty$. However

$$\varphi_1\left(\frac{t}{p}\right) - \varphi_1\left(\frac{1}{p}\right) \to 0, \quad \text{as} \, p \to +\infty$$

from condition (a) of Lemma 3.5.1'. Hence, the ratio $l(p/t)/l(p)$ tends to 1, as $p \to +\infty$. That is, $l(p)$ slowly varies as $p \to +\infty$.

Furthermore by virtue of Lemma 3.3.1

$$\frac{l(p)}{p^k} \equiv e^{-p\overline{\varphi}(p)} = \int_{0-}^{\infty} e^{-pt} \, dU(t)$$

where $U(t)$ is a non-decreasing function (which is supposed to be equal to zero for $t < 0$). Hence, by Karamata's theorem

$$U(t) \sim \frac{t^k}{\Gamma(k+1)} l\left(\frac{1}{t}\right) = \frac{t^k}{\Gamma(k+1)} \left(\frac{1}{t}\right)^k \exp\left[-\frac{1}{t}\overline{\varphi}\left(\frac{1}{t}\right)\right], t \to +0,$$

whence by virtue of Lemma 3.6.1

$$U(t) \sim \frac{e^{-kC-\varphi(t)}}{\Gamma(k+1)}, \quad t \to +0.$$

The lemma is proved. Lemma 3.6.2 obviously leads us to the following result.

Theorem 3.6.2 Suppose $\varphi(t)$ satisfies the conditions of Lemma 3.6.1. Then

$$\varepsilon(t,x) \sim \frac{e^{-kCx/(2c)}}{\Gamma\left(\frac{kx}{2c}+1\right)} \exp\left[-\varphi\left(t-\frac{x}{c}\right)\frac{x}{2c}\right], \quad \text{as } t \to \frac{x}{c}+0 \qquad (3.6.13)$$

where $C = 0.57\ldots$ is the Euler's constant.

Now, let us pass to the evaluation of $\varepsilon(t, x)$ by means of Tauberian inequalities in the manner of Lemma 3.5.3 and Theorem 3.5.2.

Lemma 3.6.3 Let

$$\varphi(t) = k \ln \frac{1}{t} \text{ for } 0 < t \le \gamma \qquad (3.6.14)$$

where $k > 0$, $\gamma > 0$, and suppose

$$a \le -t\varphi'(t) \le b, \quad \text{for } \gamma < t < t_0 \qquad (3.6.15)$$

where $0 < a \le k \le b$. Then

$$t^a \frac{e^{-kC}}{\Gamma(k+1)}\gamma^{k-a} \le U(t) = L_{p\to t}^{-1} \frac{e^{-p\overline{\varphi}(p)}}{p} \le t^b \frac{e^{-kC}}{\Gamma(k+1)\gamma^{b-k}}, \quad \text{for } 0 \le t$$

$$< t_0 \qquad (3.6.16)$$

where $C = 0.57\ldots$ is the Euler's constant, Γ the Gamma function.

Proof As we know, from Eq. (3.5.16), the function

$$u(t) = L_{p\to t}^{-1} e^{-p\overline{\varphi}(p)}$$

satisfies the Volterra equation

$$tu(t) = -\int_0^t \tau\varphi'(\tau)u(t-\tau)\mathrm{d}\tau \qquad (3.6.17)$$

Furthermore, since $0 < a \leq k \leq b$, it follows from (3.6.14, 3.6.15) that

$$0 < a \leq -t\varphi'(t) \leq b, \quad \text{for } 0 < t < t_0.$$

Hence (3.6.17) yields

$$a\int_0^t u(\tau)\mathrm{d}\tau \leq tu(t) \leq b\int_0^t u(\tau)\mathrm{d}\tau, \quad \text{for } 0 \leq t < t_0 \qquad (3.6.18)$$

Since

$$\int_0^t u(\tau)\mathrm{d}\tau = L_{p\to t}^{-1}\frac{e^{-p\bar{\varphi}(p)}}{p} \equiv U(t),$$

Eq. (3.6.18) can obviously be rewritten as

$$ad\,\ln t \leq \mathrm{d}\,\ln U(t) \leq bd\,\ln t$$

whence integrating over $\mathrm{d}t$ on (γ, t) (where $t < t_0$), one easily obtains

$$\ln\frac{t^a}{\gamma^a} \leq \ln\frac{U(t)}{U(\gamma)} \leq \ln\frac{t^b}{\gamma^b}. \qquad (3.6.19)$$

Now, note that Eq. (3.6.14) yields

$$U(\gamma) = e^{-kC}\frac{\gamma^k}{\Gamma(k+1)} \qquad (3.6.20)$$

where C is the Euler's constant. Formulas (3.6.19, 3.6.20) yield the result required.

Now, substituting in (3.6.16) t for $(t-x)/c$ and k, a, b for $kx/(2c)$, $ax/(2c)$, $bx/(2c)$, one easily gets the following result.

Theorem 3.6.3 Under the assumptions of Lemma 3.6.3

$$\left(t - \frac{x}{c}\right)^{\frac{ax}{2c}} \frac{e^{-\frac{kCx}{2c}} \gamma^{\frac{(k-a)x}{2c}}}{\Gamma\left(\frac{kx}{2c} + 1\right)}$$

$$\leq \varepsilon(t, x) \tag{3.6.21}$$

$$\leq \left(t - \frac{x}{c}\right)^{bx/(2c)} \frac{e^{-kCx/(2c)}}{\Gamma\left(\frac{kx}{2c} + 1\right) \gamma^{(b-k)x/(2c)}}, \quad \text{for } \frac{x}{c} \leq t < \frac{x}{c} + t_0$$

where $C = 0.57\ldots$ is the Euler's constant, Γ the Gamma function.

3.7 Memory Function with a Singularity Stronger than the Logarithmic One

In this section, we consider the case where the memory function $\varphi(t)$ satisfies the condition

$$\lim_{t \to +0} \frac{\varphi(t)}{\ln (1/t)} = +\infty \tag{3.7.1}$$

As we shall see below, in this case, the method of computation of the wave front asymptotics of $\varepsilon(t, x)$, based on the theory of regularly varying functions, leads to comparatively weak results. Fortunately, one can employ, the approach based on Tauberian inequalities by use of which one can evaluate the solution $\varepsilon(t, x)$.

We shall begin with the following assertion of general character.

Theorem 3.7.1 Suppose Eq. (3.7.1) holds. Then for $x > 0$, the solution $\varepsilon(t, x)$ of the problem (3.1.1, 3.1.2, and 3.1.3) is infinitely differentiable and, in particular, is infinitely differentiable in the vicinity of the wave front $t = x/c, x > 0$.

Proof As we know from Eq. (3.1.4)

$$\tilde{\varepsilon}(\lambda, x) = \frac{1}{i\lambda} \exp\left\{-i\lambda\left[1 + \frac{\tilde{\varphi}(\lambda)}{2}\right] \frac{x}{c}\right\}.$$

Furthermore, by virtue of Lemma 1.5.4, for $p > 0$

$$\text{Im}\,[\lambda\widetilde{\varphi}(\lambda)] \leq -\,\varphi\left(\frac{\pi}{2|\mu|}\right)\exp\left(-\frac{p\pi}{2|\mu|}\right), \quad \mu \neq 0,$$
$$\leq 0, \mu = 0 \tag{3.7.2}$$

Now, let us fix $p = p_0 > 0$ large enough. Then for $\lambda = \mu - ip_0$, one obtains

$$|\widetilde{\varepsilon}(\lambda,x)| = \left|\frac{1}{i\lambda}\exp\left\{-i\lambda\left[1 + \frac{\widetilde{\varphi}(\lambda)}{2}\right]\frac{x}{c}\right\}\right|$$

$$\leq \text{const}\,\exp\left[-\frac{x}{2c}\,\varphi\left(\frac{\pi}{2|\mu|}\right)\exp\left(-\frac{p_0\pi}{2|\mu|}\right)\right], \quad \mu \neq 0,$$

$$\leq \text{const}, \quad \mu = 0$$

whence by virtue of Eq. (3.7.1), it follows that for $x \neq 0$ and for $k > 0$, however large, there exists $a_k = a_k(x, p_0)$, such that

$$|\widetilde{\varepsilon}(\lambda,x)| \leq a_k(1 + |\mu|)^{-k}, \quad \lambda = \mu - ip_0. \tag{3.7.3}$$

As is well-known [6], it follows from Eq. (3.7.3) that $\varepsilon(t,x)$ is infinitely differentiable in t (for $x > 0$).

Let us pass to the demonstration of $\varepsilon(t,x)$ being infinitely differentiable in x for $x > 0$. By virtue of Eq. (3.7.2), we have for $\lambda = \mu - ip_0$,

$$\left|\frac{\partial^k\widetilde{\varepsilon}(\lambda,x)}{\partial x^k}\right| = \frac{1}{|i\lambda|}\left|-\frac{i\lambda}{c}\left[1 + \frac{\widetilde{\varphi}(\lambda)}{2}\right]\right|^k\left|\exp\left\{-i\lambda\left[1 + \frac{\widetilde{\varphi}(\lambda)}{2}\right]\frac{x}{c}\right\}\right|$$

$$\leq \frac{1}{|\lambda|}\left|\frac{\lambda}{c}\left(1 + \frac{\widetilde{\varphi}(\lambda)}{2}\right)\right|^k\exp\left[-\frac{|x|}{2c}\,\varphi\left(\frac{\pi}{2|\mu|}\right)e^{-\frac{p_0\pi}{2|\mu|}}\right], \mu \neq 0, \tag{3.7.4}$$

$$\leq \frac{1}{|\lambda|}\left|\lambda\left[1 + \frac{\widetilde{\varphi}(\lambda)}{2}\right]\right|^k, \quad \mu = 0$$

where $k = 0, 1, 2, \dots$. Hence by virtue of Eq. (3.7.1), it is clear that for $x > \delta$ (where $\delta > 0$ is however small) and for each $k = 0, 1, 2, \dots$, the function $\partial^k\widetilde{\varepsilon}(\lambda,x)/\partial x^k$ is Lebesgue integrable over $d\mu$ and the following estimate holds

$$\int_{-\infty}^{\infty}\left|\frac{\partial^k}{\partial x^k}\,\widetilde{\varepsilon}(\lambda,x)\right|d\mu \leq b_k = b_k(\delta,p_0); \lambda = \mu - ip_0. \tag{3.7.5}$$

Hence for each k

$$F_{\mu \to t}^{-1} \frac{\partial^k}{\partial x^k} \widetilde{\varepsilon}(\lambda, x) = e^{-p_0 t} \frac{\partial^k}{\partial x^k} \varepsilon(t, x)$$

is a uniformly continuous in t equicontinuously depending on x function, provided x is bounded away from zero.

But from Eq. (3.1.4), it evidently follows that the function $\partial^k \varepsilon(t, x)/\partial x^k$ considered as a distribution in t, continuously depends on the parameter x for $x > 0$. Hence by virtue of the above-mentioned follows the continuity (in the usual sense) of the function $\partial^k \varepsilon(t, x)/\partial x^k$, for $x > 0$. Since k can be chosen however large, the infinite differentiability of $\varepsilon(t, x)$, $x > 0$, is proved.

Corollary Under the condition (3.7.1), the function

$$u(t) = F_{\lambda \to t}^{-1} e^{-i\lambda \widetilde{\varphi}(\lambda)} = L_{p \to t}^{-1} e^{-p\widetilde{\varphi}(p)}$$

is infinitely differentiable on the whole of t-axis.

Note Let us adduce the following asymptotic result, which is based on a Tauberian theorem by Feller. Suppose

$$\varphi(t) = \frac{t^{-\alpha}}{\Gamma(1 - \alpha)} l\left(\frac{1}{t}\right), \quad 0 < \alpha < 1 \tag{3.7.6}$$

where $l(p)$ is a function slowly varying as $p \to +\infty$. One can easily see that in this case

$$\overline{\varphi}(p) = [1 + o(1)]p^{\alpha - 1} l(p), p \to +\infty \tag{3.7.7}$$

whence it follows that for $a > 1$

$$\frac{e^{-ap\overline{\varphi}(ap)}}{e^{-p\overline{\varphi}(p)}} \to 0, \quad \text{as } p \to +\infty \tag{3.7.8}$$

Since

$$e^{-p\overline{\varphi}(p)} = \int\limits_0^\infty e^{-pt} \, dU(t)$$

where $U(t)$ is a non-decreasing continuous function, it follows from (3.7.8) by a Tauberian Feller's theorem [2] that

$$U(t) = o\left\{\exp\left[-\frac{1}{t}\overline{\varphi}\left(\frac{1}{t}\right)\right]\right\}, \text{ as } t \to +0.$$

Substituting Eq. (3.7.7) into the last relation, we finally obtain

$$U(t) \equiv L_{p\to t}^{-1}\frac{e^{-p\overline{\varphi}(p)}}{p} = o\left(\exp\left\{-[1+o(1)]\,t^{-\alpha}l\left(\frac{1}{t}\right)\right\}\right)$$

$$= o\left\{e^{-[1+o(1)]\Gamma(1-\alpha)\varphi(t)}\right\}, \quad t \to +0,$$

(3.7.9)

Therefore, one easily obtains the following result

$$\varepsilon(t,x) = o\left(\exp\left\{-[1+o(1)]\Gamma(1-\alpha)\varphi\left(t-\frac{x}{c}\right)\frac{x}{2c}\right\}\right), \text{ as } t \to \frac{x}{c}+0. \quad (3.7.10)$$

Now, let us pass to the study of $\varepsilon(t,x)$ by the method of Tauberian inequalities.

Lemma 3.7.1 Suppose $\varphi(t)$ satisfies the condition (3.7.1), and let $-t\varphi'(t)$ be non-increasing for $t > 0$. Then

$$0 \le L_{p\to t}^{-1}\frac{e^{-p\overline{\varphi}(p)}}{p} \le e^{-\varphi(t)}, \quad \text{for } t \ge 0. \quad (3.7.11)$$

Example

$$0 \le L_{p\to t}^{-1}\frac{1}{p}\exp\left(-\sum_j c_j p^{\alpha_j}\right) \le \exp\left[-\sum_j c_j\frac{t^{-\alpha_j}}{\Gamma(1-\alpha_j)}\right], \text{ for } t \ge 0 \quad (3.7.12)$$

Here $c_j > 0$, $0 < \alpha_j < 1$.

Proof of the Lemma As we know, the function

$$u(t) = L_{p\to t}^{-1}e^{-p\overline{\varphi}(p)}$$

satisfies the equation

$$tu(t) = -\int_0^t \tau\varphi'(\tau)u(t-\tau)\,d\tau \quad (3.7.13)$$

(see Sect.3.5). Furthermore, $u(t) \ge 0$, for $t > 0$ (see Corollary to Lemma 3.3.5) and $\varphi'(t) \le 0$, for $t > 0$ by virtue of property (b) from Sect. 3.1.1. Hence, the integrand in (3.7.13) does not change its sign. So, by virtue of $-t\varphi'(t)$ being non-increasing for $t > 0$, one obviously gets from (3.7.13)

$$t u(t) \geq -t\varphi'(t) \int_0^t u(\tau)\, d\tau, \quad t > 0,$$

whence

$$\frac{d}{dt}[\varphi(t) + \ln U(t)] \geq 0 \tag{3.7.14}$$

where

$$U(t) = \int_0^t u(\tau)\, d\tau = L_{p \to t}^{-1} \frac{e^{-p\bar{\varphi}(p)}}{p}.$$

Thus, it follows from Eq. (3.7.14) that

$$\int_t^\infty d[\varphi(\tau) + \ln U(\tau)] \geq 0, t > 0. \tag{3.7.15}$$

Note now that by virtue of Lemma 3.3.2

$$\varphi(\infty) + \ln U(\infty) \leq 0$$

Hence (3.7.15) yields

$$\varphi(t) + \ln U(t) \leq 0,$$

That is

$$U(t) \leq e^{-\varphi(t)},$$

which proves the right-hand inequality in Eq. (3.7.11). The left-hand inequality was established in Sect. 3.3.

Now, let us pass to the evaluation of $u(t)$.

Lemma 3.7.2 Suppose (3.7.1) holds, the function $-t\varphi'(t)$ is non-increasing for $t > 0$, and the function $\varphi(t) - \ln(1/t)$ is non-increasing on $(0, t_0)$ and does not possess this property on any larger interval. Then

$$0 \leq u(t) = L_{p \to t}^{-1} e^{-p\bar{\varphi}(p)} \leq \frac{e^{-C - \varphi(t)}}{t}, \quad \text{for } 0 \leq t < t_0 \tag{3.7.16}$$

and $u(t)$ is non-decreasing for $0 \leq t < t_0$. Here $C = 0.57\ldots$ is the Euler's constant.

Proof Since

$$L_{t \to p} \ln \frac{1}{t} = \frac{C + \ln p}{p},$$

we have

$$e^{-p\overline{\varphi}(p)} = e^{-C} \frac{1}{p} \exp \left\{ -pL_{t \to p} \left[\varphi(t) - \ln \left(\frac{1}{t} \right) \right] \right\}. \tag{3.7.17}$$

Let

$$
\begin{aligned}
\psi(t) &= \varphi(t) - \ln \frac{1}{t}, \quad \text{for } 0 < t < t_0, \\
&= \varphi(t_0) - \ln \frac{1}{t_0}, \quad \text{for } t \geq t_0.
\end{aligned}
\tag{3.7.18}
$$

Then by virtue of Lemma 3.1.1

$$L_{p \to t}^{-1} \frac{1}{p} \exp \left\{ -pL_{t \to p} \left[\varphi(t) - \ln \left(\frac{1}{t} \right) \right] \right\} = L_{p \to t}^{-1} \frac{e^{-p\overline{\psi}(p)}}{p}, \text{ for } 0 \leq t < t_0. \tag{3.7.19}$$

Furthermore, in the manner of Lemma 3.3.1 and Theorem 3.7.1, one can easily establish that

$$L_{p \to t}^{-1} \frac{e^{-p\overline{\psi}(p)}}{p}$$

is a non-decreasing, non-negative, infinitely differentiable function. Moreover, since $-t\varphi'(t)$ is non-increasing for $t > 0$, it follows from (3.7.18) that $-t\psi'(t)$ is also non-increasing for $t > 0$. Hence in the manner of Lemma 3.7.1, one obtains

$$0 \leq L_{p \to t}^{-1} \frac{e^{-p\overline{\psi}(p)}}{p} \leq e^{-\psi(t)}, \quad \text{for } t \geq 0 \tag{3.7.20}$$

whence by virtue of (3.7.17, 3.7.18, and 3.7.19), it follows that for $0 \leq t < t_0$,

$$L_{p \to t}^{-1} e^{-p\overline{\varphi}(p)} \leq e^{-C} e^{-\psi(t)} = e^{-C} \frac{e^{-\varphi(t)}}{t},$$

which gives the right-hand inequality in Eq. (3.7.16). The left-hand inequality was established in Sect. 3.3.

Thus we arrive at the following result.

Theorem 3.7.2

(a) Suppose Eq. (3.7.1) holds and the function $-t\varphi'(t)$ is non-increasing for $t > 0$.
Then

$$0 \le \varepsilon(t,x) \le \exp\left[-\varphi\left(t - \frac{x}{c}\right)\frac{x}{2c}\right], \text{ for } t \ge \frac{x}{c}. \qquad (3.7.21)$$

(b) Moreover, suppose also that for each $x > 0$

$$\varphi(t)\frac{x}{2c} - \ln\frac{1}{t}$$

is a non-increasing in t function on the interval $(0, t_0(x))$ and does not possess this
property on any larger interval. Then

$$0 \le \varepsilon_t'(t,x) \le \frac{e^{-C}}{t - (x/c)}\, e^{-\varphi\left(t-\frac{x}{c}\right)\frac{x}{2c}}, \text{ for } \frac{x}{c} \le t < \frac{x}{c} + t_0(x) \qquad (3.7.22)$$

Here $C = 0.57...$ is the Euler's constant.

3.8   The Power Memory Function

Let

$$\varphi(t) = \frac{k}{\Gamma(1 - \alpha)}\, t^{-\alpha}, \text{ for } t > 0 \qquad (3.8.1)$$

where $k > 0$, $0 < \alpha < 1$,

$$u_\alpha(t) = L_{p \to t}^{-1} e^{-p^\alpha}. \qquad (3.8.2)$$

By virtue of Corollary to Theorem 3.7.1, $u_\alpha(t)$ is infinitely differentiable on the
whole of t-axis. Furthermore, from Corollary to Lemma 3.3.5, it follows that $u_\alpha(t) \ge 0$
(for the first time this fact was established in [7]).

Let us adduce the following result concerning the behaviour of $u_\alpha(t)$.

Lemma 3.8.1[7]

$$u_\alpha(t) = \frac{1}{\pi} \sum_{n=1}^{\infty} (-1)^{n-1} \frac{\sin(\pi n \alpha)}{n!} \frac{\Gamma(n\alpha + 1)}{t^{n\alpha+1}}, \quad \text{for } t > 0 \qquad (3.8.3)$$

Γ is the Gamma function.

The expansion (3.8.3) converges non-uniformly for small $t > 0$. So, of interest is the following result giving the asymptotic representation for $u_\alpha(t)$, as $t \to +0$ [8]

$$u_\alpha(t) \sim \frac{\alpha^{1/2(1-\alpha)}}{\sqrt{2\pi(1-\alpha)}}\, t^{-1-\frac{\alpha}{2(1-\alpha)}} \exp\left[-(1-\alpha)\alpha^{\alpha/(1-\alpha)} t^{\alpha/(1-\alpha)}\right]. \qquad (3.8.4)$$

Note that in case of $\alpha = 1/2$ (3.8.4) passes into a well-known precise equality

$$u_{1/2}(t) = \frac{1}{2\pi^{1/2}t^{3/2}}\, \exp\left(-\frac{1}{4t}\right). \qquad (3.8.5)$$

The following simple theorem immediately follows from formula

$$\varepsilon(t,x) = L^{-1}_{p \to t-(x/c)}\, \frac{e^{-kp^\alpha x/(2c)}}{p}.$$

Theorem 3.8.1 [1] Under the (3.8.1)

$$\varepsilon(t,x) = \int\limits_0^{\frac{t-x/c}{(kx/2c)^{1/\alpha}}} u_\alpha(\xi)\, \mathrm{d}\xi \qquad (3.8.6)$$

Note: It is known (see [9]) that $u_\alpha(t)$ has a unique maximum on the semi-axis $t \geq 0$. Let us show that this maximum is achieved in some point t_m of the segment $[t_0, t_1]$, where

$$t_0 = \left[\frac{\alpha}{\Gamma(1-\alpha)}\right]^{1/\alpha}, \qquad t_1 = \left[\frac{\alpha}{\Gamma(2-\alpha)}\right]^{1/\alpha} \qquad (3.8.7)$$

At first, let us prove that the maximum of $u_\alpha(t)$ is located to the right of t_0. In fact, one can easily see that

$$\frac{t^{-\alpha}}{\Gamma(1-\alpha)} - \ln\frac{1}{t}$$

is monotone decreasing on the interval $(0, t_0)$, where the value of t_0 is given by (3.8.7). Hence, by Lemma 3.7.2, it follows that $t_m > t_0$.

Now, let us prove that $t_m < t_1$. We have

$$tu_\alpha(t) = \frac{\alpha}{\Gamma(1-\alpha)} \int\limits_0^t (t-\tau)^{-\alpha} u_\alpha(\tau)\,d\tau. \tag{3.8.8}$$

This equality is a special case of (3.5.16). It can also be checked by direct application of the Laplace transform. Since $u_\alpha(t) \geq 0$, it follows from Eq. (3.8.8) and monotone increase of $u_\alpha(t)$ on $(0, t_m)$ that

$$t_m u_\alpha(t_m) \leq \frac{\alpha}{\Gamma(1-\alpha)} u_\alpha(t_m) \int\limits_0^{t_m} (t_m - \tau)^{-\alpha}\,d\tau$$

$$= \frac{\alpha}{\Gamma(1-\alpha)} u_\alpha(t_m) \frac{(t_m)^{1-\alpha}}{1-\alpha}$$

whence

$$t_m \leq \left[\frac{\alpha}{\Gamma(2-\alpha)}\right]^{1/\alpha} = t_1.$$

3.9   Application of the Nonlinear Laplace Transform to Calculating the Wave Front Velocity in an Inhomogeneous Hereditary Rod

3.9.1   Formulation of the Problem

Consider an inhomogeneous rod located on the semi-axis $x \geq 0$ and having the density $\rho = \rho(x)$. We suppose that in the rod under consideration stress σ and deformation ε are related by

$$\sigma = A(x)\left[\varepsilon - R(t) * \varepsilon\right] \tag{3.9.1}$$

or which is the same,

$$\varepsilon = \frac{1}{A(x)}\left[\sigma + \Lambda(t) * \sigma\right] \tag{3.9.1''}$$

Here $A(x)$ is the instantaneous module of elasticity, $R(t)$ the relaxation kernel, and $\Lambda(t)$ the creep kernel. The relation between $R(t)$ and $\Lambda(t)$ is given by the formula (see Sect. 1.1.1)

$$\frac{1}{1 - R(t)*} = 1 + \Lambda(t)*$$

As everywhere in this chapter

$$\Lambda(t) = \varphi(t) + \frac{1}{4}\varphi(t) * \varphi(t) \tag{3.9.2}$$

where $\varphi(t)$ satisfies conditions (a)–(e) from Sect. 3.1.

This section is devoted to the problem of calculating the longitudinal wave front velocity in the rod under consideration.

Let us write out equations of the rod motion:

$$\frac{1}{\rho(x)}\frac{\partial \sigma}{\partial x} = \frac{\partial v}{\partial t}; \frac{\partial v}{\partial x} = \frac{\partial \varepsilon}{\partial t} \tag{3.9.3}$$

where v is the velocity of the rod element. Eliminating v from the system (3.9.3) and substituting σ for its expression (3.9.1), we arrive at the following wave equation with memory and variable coefficients:

$$\frac{\partial^2 \varepsilon}{\partial t^2} - \frac{\partial}{\partial x}\frac{1}{\rho(x)}\frac{\partial}{\partial x}\{A(x)[\varepsilon - R(t) * \varepsilon]\} = 0$$

or which is the same,

$$[1 + \Lambda(t)*]\frac{\partial^2 \varepsilon}{\partial t^2} - \frac{\partial}{\partial x}\frac{1}{\rho(x)}\frac{\partial}{\partial x}[A(x)\varepsilon] = 0 \tag{3.9.4}$$

Let us set the following problem for Eq. (3.9.4):

$$\varepsilon(t, x) = 0 \text{ for } x > 0, t \leq 0$$
$$\varepsilon(t, 0) = \varepsilon_0(t) \tag{3.9.5}$$

where $\varepsilon_0(t) = 0$, for $t < 0$ and $\varepsilon_0(+0) > 0$. In what follows, it will be convenient for us to take $\varepsilon_0(t)$ as the function

$$\varepsilon_0(t) = L_{p \to t}^{-1}\frac{1}{P\sqrt{1 + \overline{A}(p)}} = L_{p \to t}^{-1}\frac{1}{p\left[1 + \frac{\overline{\varphi}(p)}{2}\right]} \tag{3.9.6}$$

It is easy to see that when calculating the wave front velocity, we can restrict ourselves to the special case (3.9.6) without loss of generality.

Note For completeness of exposition, let us recall the classical method of calculating the wave front velocity in the problem under consideration in case of regular function of memory. In this case, the line of wave front, $t = t(x)$, issues out of the origin of coordinates. In front of the wave front $\varepsilon = \sigma = 0$, while across the line $t = t$ (x), ε and σ suffer finite jumps (see [10]).

Let us introduce the following notation for the jump of $f(t)$ across the wave front $t = t(x)$:

$$[\![f]\!] \equiv f(t(x) + 0, x) - f(t(x) - 0, x).$$

As is well known, the equations of motion (3.9.3) must be supplemented with the following conditions on the wave front

$$\frac{1}{\rho(x)} [\![\sigma]\!] = -c[\![v]\!]; \ [\![v]\!] = -c[\![\varepsilon]\!] \tag{3.9.7}$$

where c is the front velocity. Besides, by virtue of continuity of the convolution R $(t) * \varepsilon$, it follows from Eq. (3.9.1) that

$$[\![\sigma]\!] = A(x)[\![\varepsilon]\!]. \tag{3.9.8}$$

Eliminating $[\![\sigma]\!]$ and $[\![v]\!]$ from (12.6, 12.7), we have

$$\frac{A(x)}{\rho(x)} [\![\varepsilon]\!] = c^2 [\![\varepsilon]\!]. \tag{3.9.9}$$

Now, by cancellation by the nonzero jump $[\![\varepsilon]\!]$, we obtain the wave front velocity,

$$c = \sqrt{\frac{A(x)}{\rho(x)}}. \tag{3.9.10}$$

(We have chosen the sign "plus" in front of the square root because, in the problem under consideration, the wave evidently propagates to the right.)

Now, note that in case of singular memory $[\![\sigma]\!] = [\![\varepsilon]\!] = 0$ (see Sect. 3.3). Therefore, the relation (3.9.9) takes on the form

$$\frac{A(x)}{\rho(x)} \cdot 0 = c^2 \cdot 0$$

from which one cannot determine the wave front velocity.

Surely, by analogy with the homogeneous case (see Sect. 3.1), it is natural to expect that for an inhomogeneous rod with singular memory the wave front velocity will be equal to the instantaneous elastic one, that is, (3.9.10). But how should one prove this fact in absence of an explicit formula for $\bar{\varepsilon}(p, x)$?

3.9.2 Solution of the Problem

Let us apply the Laplace transform $L_{t \to p}$ to the problem (3.9.4, 3.9.5, and 3.9.6). On account of Eq. (3.9.2), we have

$$p^2 \left[1 + \frac{\overline{\varphi}(p)}{2} \right]^2 \overline{\varepsilon}(p, x) - \frac{d}{dx} \frac{1}{\rho(x)} \frac{d}{dx} [A(x) \overline{\varepsilon}(p, x)] = 0 \qquad (3.9.11)$$

and

$$\overline{\varepsilon}(p, 0) = \frac{1}{p \left[1 + \frac{\overline{\varphi}(p)}{2} \right]}. \qquad (3.9.12)$$

Since Eq. (3.9.11), generally speaking, cannot be integrated in quadratures, at the first glance, it is impossible to make use of the techniques developed above.

Our main idea is to introduce a nonlinear change of the Laplace variable by formula

$$q = p \left[1 + \frac{\overline{\varphi}(p)}{2} \right]. \qquad (3.9.13)$$

Then (3.9.11, 3.9.12) takes the form

$$q^2 \overline{w}(q, x) - \frac{d}{dx} \frac{1}{\rho(x)} \frac{d}{dx} [A(x) \overline{w}] = 0, \qquad (3.9.14)$$

$$\overline{w}(q, 0) = \frac{1}{q}$$

where it is denoted

$$\overline{w}(q, x) \equiv \overline{\varepsilon}(p(q), x). \qquad (3.9.15)$$

However, it is easy to see that Eq. (3.9.14) coincides with the Laplace transform $L_{t \to q}$ of the following problem for a purely differential equation with variable coefficients

$$\frac{\partial^2 w}{\partial t^2} - \frac{\partial}{\partial x}\frac{1}{\rho(x)}\frac{\partial}{\partial x}[A(x)w] = 0;$$

$$w = \frac{\partial w}{\partial t} = 0, \quad \text{for } x > 0, t = 0; \tag{3.9.16}$$

$$w(t,0) = \begin{cases} 1 & \text{for } t > 0 \\ 0 & \text{for } t < 0 \end{cases},$$

We suppose $A(x)$ and $\rho(x)$ to be such that Eq. (3.9.16) has a solution $w(t, x)$ smooth behind the wave front

$$t = t(x) = \int\limits_0^x \left[\frac{A(x)}{\rho(x)}\right]^{-1/2} dx \tag{3.9.17}$$

and satisfying the estimate

$$w'_t = o(e^{Mt}), M > 0,$$

for large t. As is known, on the wave front the solution $w(t, x)$ suffers the jump

$$[\![w]\!] = \left[\frac{\rho(x)}{\rho(0)}\right]^{1/4}\left[\frac{A(0)}{A(x)}\right]^{3/4} > 0.$$

However, from the above-mentioned equation, it follows that

$$|\overline{w}(q,x)| \le \frac{\text{const}}{|q|} e^{-t(x)\,\text{Re}\,q}, \quad \text{for Re } q > M > 0 \tag{3.9.18}$$

where const >0. Besides, for real q, the following estimate holds

$$\overline{w}(q,x) \sim \frac{e^{-qt(x)}}{q}[\![w]\!]; q \to +\infty \tag{3.9.19}$$

(see [11]).

Now, let us return to the variable p by formula (3.9.13). Then (3.9.18) yields

$$|\overline{\varepsilon}(p,x)| \equiv |\overline{w}(q(p),x)| \le \frac{\text{const}}{\left|p\left[1+\frac{\overline{\varphi}(p)}{2}\right]\right|} \exp\left(-t(x)\,\text{Re}\left\{p\left[1+\frac{\overline{\varphi}(p)}{2}\right]\right\}\right) \tag{3.9.20}$$

for

$$\text{Re}\left\{p\left[1+\frac{\overline{\varphi}(p)}{2}\right]\right\}>M$$

But by virtue of Lemma 1.5.4

$$\text{Re}\,[p\overline{\varphi}(p)]\geq 0,\quad\text{for Re}\,p>0. \tag{3.9.21}$$

Clearly (3.9.20, 3.9.21) yield

$$|\overline{\varepsilon}(p,x)|\leq\frac{\text{const}}{\left|p\left[1+\frac{\overline{\varphi}(p)}{2}\right]\right|}\,e^{-t(x)\,\text{Re}\,p},\text{for Re}\,p>M. \tag{3.9.22}$$

Hence by Theorem 1.3.1, it follows that $\varepsilon(t,x)=0$ for $t<t(x)$. (This fact was established in [12].)

Finally, let us demonstrate that $t=t(x)$ is the precise equation of the wave front for $\varepsilon(t,x)$. Suppose the contrary. Then for some $x>0$, there must exist $\delta_1>0$ such that $\varepsilon(t,x)=0$ for $t\leq t(x)+\delta_1$. Then by Theorem 1.3.1, we must have the following estimate

$$|\overline{\varepsilon}(p,x)|\leq C_1\,(1+|p|)^{\nu_1}\,e^{-[t(x)+\delta_1]\,\text{Re}\,p},\quad\text{for Re}\,p>M_1$$

where C_1, ν_1, M_1 are some positive constants. But, the last estimate contradicts Eq. (3.9.19).

The contradiction obtained demonstrates that in inhomogeneous rods with singular memory wave fronts propagate precisely at instantaneous elastic velocity.

Note that we have obtained the solution of our problem by reducing it to a classical one by applying the nonlinear Laplace transform

$$\varepsilon(t,x)=L_{p\to t}^{-1}\,L_{t\to q(p)}\,w(t,x). \tag{3.9.23}$$

3.10 Asymptotic Generalization of the Cagniard–de Hoop Method: The Case of a Line Source

As is well known, the classical two-dimensional Cagniard–de Hoop method (see [13, 14]) enables us to analytically solve problems about wave propagation generated by a line source in linear elastic layered media (see [13]). In this section, we give a generalization of the mentioned method to the hereditary case. For simplicity, we demonstrate the idea of our generalization in solving the simplest problem about shear wave propagation from a line source in an isotropic homogeneous hereditary space. It should be noted that our exposition is not mathematically rigorous and presents a refined variant of [15].

So, let us consider an isotropic homogeneous hereditary elastic *xyz*-space of density $\rho = \text{const} > 0$. As we know, equations of motion of the medium under consideration have the form (see 1.1.10)

$$\rho\frac{\partial^2 \vec{u}}{\partial t^2} = \{\lambda[1 - q(t)*] + 2\mu[1 - h(t)*]\}\nabla\left(\nabla \times \vec{u}\right)$$
$$-\mu[1 - h(t)*]\nabla \times \left(\nabla \times \vec{u}\right) + \vec{f} \tag{3.10.1}$$

Here $\vec{u} = (u_x, u_y, u_z)$ is the vector of displacement, $\vec{f} = (f_x, f_y, f_z)$ the body force; $\lambda > 0$, and $\mu > 0$ are the instantaneous constants of Lamé, and $q(t)$ and $h(t)$ are the corresponding relaxation kernels. Let

$$\vec{f} = (0, A\delta(x)\,\delta(z)\,\delta(t), 0) \tag{3.10.2}$$

where $A = \text{const} \neq 0$, and suppose

$$\vec{u} = 0, \quad \text{for } t < 0. \tag{3.10.3}$$

Clearly, in such a setting u_x and u_z will remain identically equal to zero. For u_y (3.10.1) yields

$$\frac{\partial^2 u_y}{\partial t^2} - \beta_0^2[1 - h(t)*]\left(\frac{\partial^2 u_y}{\partial x^2} + \frac{\partial^2 u_y}{\partial z^2}\right) = \frac{A}{\rho}\,\delta(t)\,\delta(x)\,\delta(z) \tag{3.10.4}$$

where

$$\beta_0 = \left(\frac{\mu}{\rho}\right)^{1/2} \tag{3.10.5}$$

Let $\Lambda(t)$ be the creep kernel corresponding to the relaxation kernel $h(t)$

$$\frac{1}{1 - h(t)*} = 1 + \Lambda(t)*$$

Then Eq. (3.10.4) can evidently be rewritten in the form

$$[1+\Lambda(t)*]\frac{\partial^2 u_y}{\partial t^2} - \beta_0^2\left(\frac{\partial^2 u_y}{\partial x^2} + \frac{\partial^2 u_y}{\partial z^2}\right) = \frac{A}{\rho}[1+\Lambda(t)*]\,\delta(t)\,\delta(x)\,\delta(z) \quad (3.10.6)$$

Just like Eq. (3.1.3), we suppose

$$\Lambda(t) = \varphi(t) + \frac{1}{4}\,\varphi(t)*\varphi(t) \tag{3.10.7}$$

where $\varphi(t)$ satisfies conditions (a)–(e) from Sect. 3.1.

In the manner of Theorem 2.4.1, one can demonstrate that under the condition (3.10.7), the operator in (3.10.6) is hyperbolic and describes wave propagation at speed β_0.

Our purpose is to derive, in the vicinity of the wave front, an asymptotic formula for the solution u_y of (3.10.3, 3.10.6, 3.10.7) without applying any integral transformation with respect to z. We leave it for the reader to show that our approach really proves to be applicable for solving wave problems in layered (with respect to z) hereditary media.

Let us apply the Laplace transform $L_{t\to p}$ and the Fourier transform $F_{x\to\xi}$ to (3.10.6, 3.10.7). After solving the resulting ordinary differential equation (with respect to z), one easily obtains on account of (3.10.3):

$$F_{x\to\xi}L_{t\to p}u_y = \frac{A}{2\rho\beta^2(p)n}e^{-n|z|} \tag{3.10.8}$$

where it is denoted as

$$n \equiv \left[\xi^2 + \frac{p^2}{\beta^2(p)}\right]^{1/2},\, \beta(p) \equiv \frac{\beta_0}{1+\frac{\overline{\varphi}(p)}{2}}. \tag{3.10.9}$$

Note Here $n \geq 0$ for ξ and p real. Below, we extend the definition of n to the case of complex ξ by using the restriction Re $n > 0$. On applying the inverse Fourier transform $F_{\xi\to x}^{-1}$ to (3.10.8), we have

$$\bar{u}_y(p, x, z) = \frac{A}{4\pi\rho\beta^2(p)}\int_{-\infty}^{\infty}\frac{e^{i\xi x - n|z|}}{n}\,d\xi. \tag{3.10.10}$$

Furthermore, just like in the classical Cagniard–de Hoop method, we introduce the *ray parameter*, s, by formula

$$\xi = isp. \tag{3.10.11}$$

From (3.10.9, 3.10.10, and 3.10.11), it follows that

$$\bar{u}_y(p,x,z) = -\frac{Ai}{4\pi\rho\beta^2(p)} \int_{-i\infty}^{i\infty} \frac{e^{-p(sx+\eta|z|)}}{\eta} \, ds$$

where $\eta \equiv [\beta^{-2}(p) - s^2]^{1/2}$, $\mathrm{Re}\, \eta > 0$. As in [13], from the last formula, one easily obtains

$$\bar{u}_y(p,x,z) = \frac{A}{2\pi\rho\beta^2(p)} \, \mathrm{Im} \int_0^{i\infty} \frac{e^{-p(sx+\eta|z|)}}{\eta} \, ds. \tag{3.10.12}$$

Now, we fix some $p > 0$, and some real x and z, and introduce the time-like variable $\tau = \tau(s)$ by formula

$$\tau = sx + \left[\beta^{-2}(p) - s^2\right]^{1/2} |z|. \tag{3.10.13}$$

In doing this, we suppose τ to run over the interval $0 < \tau < \infty$.
Let $H \equiv (x^2 + z^2)^{1/2}$. Then solving Eq. (3.10.13) for s gives, in particular, the root

$$s = \frac{x\tau - |z|\left[\frac{H^2}{\beta^2(p)} - \tau^2\right]^{1/2}}{H^2}, \quad \text{for } 0 < \tau \le \frac{H}{\beta(p)},$$

$$= \frac{x\tau + i|z|\left[\tau^2 - \frac{H^2}{\beta^2(p)}\right]^{1/2}}{H^2}, \quad \text{for } \tau > \frac{H}{\beta(p)}. \tag{3.10.14}$$

The equalities (3.10.14) determine the generalized Cagniard's contour $c_p : s = s$ (τ). Furthermore, just like in [13], (3.10.12, 3.10.13, 3.10.14) yield

$$\bar{u}_y(p,x,z) = \frac{A}{2\pi\rho\beta^2(p)} \, \mathrm{Im} \int_{C_p} \frac{e^{-p(sx+\eta|z|)}}{\eta} \, ds.$$

The part of C_p between $\tau = 0$ and $\tau = H/\beta(p)$ gives no contribution into the last expression for $\bar{u}_y(p,x,z)$, since the integrand is real along it. Besides, we have

$$\frac{ds}{d\tau} = \frac{i\eta}{\left[\tau^2 - \frac{H^2}{\beta^2(p)}\right]^{1/2}}$$

Along C_p, for $\tau > H/\beta(p)$. Therefore, on account of (3.10.14), the previous expression for $\bar{u}_y$ can be rewritten as

$$\bar{u}_y(p,x,z) = \frac{A}{2\pi\rho\beta^2(p)} \int\limits_{H/\beta(p)}^{\infty} \frac{e^{-p\tau}\,d\tau}{\left[\tau^2 - \frac{H^2}{\beta^2(p)}\right]^{1/2}}. \tag{3.10.15}$$

Now, making a change

$$\tau \to \tau\frac{\beta_0}{\beta(p)}$$

in (3.10.15) and taking into account (3.10.9), one easily obtains

$$\bar{u}_p(p,x,z) = \frac{A}{2\pi\rho\beta^2(p)} \int\limits_{H/\beta_0}^{\infty} \frac{\exp\left\{-p\left[1+\frac{\bar{\varphi}(p)}{2}\right]\tau\right\}}{\left(\tau^2 - \frac{H^2}{\beta_0^2}\right)^{1/2}}\,d\tau. \tag{3.10.15'}$$

Clearly, the integral on the right presents the nonlinear Laplace transform.

Our idea is to qualitatively describe the behaviour of $u_y(t,x,z)$ near the wave front $t = H/\beta_0$ by means of asymptotics of $\bar{u}_y(p,x,z)$ for large $p > 0$.

It is clear that for large $p > 0$, only a small part of the integration path

$$\frac{H}{\beta_0} < \tau < \frac{H}{\beta_0} + \delta,$$

gives the main contribution into the integral (3.10, 3.15'). Along the mentioned part of the integration path, one can consider

$$\left(\tau^2 - \frac{H^2}{\beta_0^2}\right)^{1/2} = \left(\tau - \frac{H}{\beta_0}\right)^{1/2}\left(\tau + \frac{H}{\beta_0}\right)^{1/2} \sim \left(\frac{2H}{\beta_0}\right)^{1/2}\left(\tau - \frac{H}{\beta_0}\right)^{1/2}.$$

Hence for $p \to +\infty$,

$$\bar{u}_y(p,x,z) \sim \frac{A}{2\pi\rho\beta^2(p)\left(\frac{2H}{\beta_0}\right)^{1/2}} \int\limits_{H/\beta_0}^{\infty} \frac{\exp\left\{-p\left[1+\frac{\bar{\varphi}(p)}{2}\right]\tau\right\}}{\left(\tau - \frac{H}{\beta_0}\right)^{1/2}}\,d\tau.$$

Letting $\tau_1 = \tau - H/\beta_0$ in the last integral and integrating, one easily obtains for $p \to +\infty$,

$$\bar{u}_y(p, x, z) \sim \frac{A\Gamma\left(\frac{1}{2}\right)}{2\pi\rho\beta^{3/2}(2H)^{1/2}} \frac{\exp\left\{-p\left[1 + \frac{\overline{\varphi}(p)}{2}\right]\frac{H}{\beta_0}\right\}}{p^{1/2}} \tag{3.10.16}$$

Now, to be specific, suppose

$$\varphi(t) = \frac{kt^{-\alpha}}{\Gamma(1-\alpha)}, \quad \text{for } t > 0 \tag{3.10.17}$$

where $k > 0$, $0 < \alpha < 1$. Then $\overline{\varphi}(p) = kp^{\alpha-1}$ and (3.10.16) assumes the form

$$\bar{u}_y(p, x, z) \sim \frac{A\Gamma(1/2)}{2\pi\rho\beta_0^{3/2}(2H)^{1/2}} \exp\left(-\frac{pH}{\beta_0}\right) \frac{\exp\left(-\frac{p^\alpha kH}{2\beta_0}\right)}{p^{1/2}}, \tag{3.10.18}$$

for $p \to +\infty$. The first multiplier on the right-hand side of (3.10.18) is independent of p. The second multiplier corresponds to the time shift $t \to t - H/\beta_0$. Therefore, it suffices to find the asymptotics of the inverse Laplace transform of the third multiplier (as $t \to +0$). Making use of technique of the method of the steepest descent [8], we obtain

$$L_{p\to t}^{-1} \frac{\exp\left(-\frac{p^\alpha kH}{2\beta_0}\right)}{p^{1/2}} \sim \frac{\left(\frac{Hk\alpha}{2\beta_0}\right)^{\frac{1}{2(1-\alpha)}}}{[2\pi(1-\alpha)]^{1/2}} t^{-\frac{1}{2(1-\alpha)}} \tag{3.10.19}$$

$$\times \exp\left\{-\left(\frac{Hk}{2\beta_0}\right)^{\frac{1}{1-\alpha}} \alpha^{\frac{\alpha}{1-\alpha}}(1-\alpha) t^{-\frac{\alpha}{1-\alpha}}\right\},$$

for $t \to +0$. Therefore, neglecting the difference between the two sides of (3.10.18), we can expect that

$$u_y(t, x, z) \sim \text{const}_1 \frac{\alpha}{H^{2(1-\alpha)}} \left(t - \frac{H}{\beta_0}\right)^{-\frac{1}{2(1-\alpha)}} \tag{3.10.20}$$

$$\times \exp\left[-\text{const}_2 H^{\frac{1}{1-\alpha}} \left(t - \frac{H}{\beta_0}\right)^{-\frac{\alpha}{1-\alpha}}\right],$$

for $t \to (H/\beta_0) + 0$, where

$$\text{const}_1 = \frac{A(k\alpha)^{\frac{1}{2(1-\alpha)}}}{\pi\rho(2\beta_0)^{\frac{3}{2}+\frac{1}{2(1-\alpha)}}\sqrt{2(1-\alpha)}},$$

and

$$\text{const}_2 = \left(\frac{k}{2\beta_0}\right)^{\frac{1}{1-a}} \alpha^{\frac{a}{1-a}}(1-\alpha).$$

Note By applying $L_{p\to t}^{-1}$ (3.10.15'), one easily obtains the following exact formula

$$u_y(t,x,z) = \frac{A}{2\pi\mu} \, [1+\Lambda(t)*] \int\limits_{H/\beta_0}^{\infty} \frac{L_{p\to t-\tau}^{-1} \, e^{-p\bar{\varphi}(p)\,(\tau/2)}}{\left(\tau^2 - \frac{H^2}{\beta_0^2}\right)^{1/2}} \, d\tau$$

$$\equiv \frac{A}{2\pi\mu} \, [1+\Lambda(t)*] \int\limits_{H/\beta_0}^{t} \frac{L_{p\to t-\tau}^{-1} \, e^{-p\bar{\varphi}(p)\,(\tau/2)}}{\left(\tau^2 - \frac{H^2}{\beta_0^2}\right)^{1/2}} \, d\tau. \tag{3.10.21}$$

Since

$$L_{p\to t-\tau}^{-1} e^{-p\bar{\varphi}(p)(\tau/2)} = 0, \quad \text{for } t < \tau,$$

the integral on the right-hand side of Eq. (3.10.21) vanishes for $t < H/\beta_0$. Therefore, Eq. (3.10.21) can be rewritten as

$$u_y(t,x,z) = \frac{A}{2\pi\mu} \, [1+\Lambda(t)*] \left[\Theta\left(t - \frac{H}{\beta_0}\right) \int\limits_{H/\beta_0}^{t} \frac{L_{p\to t-\tau}^{-1} \, e^{-p\bar{\varphi}(p)\left(\frac{\tau}{2}\right)}}{\left(\tau^2 - \frac{H^2}{\beta_0^2}\right)^{1/2}} \, d\tau \right]. \tag{3.10.22}$$

In case where Eq. (3.8.1) takes place, Eq. (3.10.22) obviously yields

$$u_y(t,x,z) = \frac{A}{2\pi\mu} \, [1+\Lambda(t)*] \left\{ \Theta\left(t - \frac{H}{\beta_0}\right) \int\limits_{H/\beta_0}^{t} \frac{u_a\left(\frac{t-\tau}{(k\tau/2)^{1/a}}\right)}{\left(\tau^2 - \frac{H^2}{\beta_0^2}\right)^{1/2}} \, d\tau \right\}. \tag{3.10.23}$$

Here
$$u_\alpha(t) \equiv L_{p\to t}^{-1} e^{-p^\alpha} \quad (\text{see Sect.3.8}).$$

3.11 Asymptotic Generalization of the Cagniard–de Hoop Method: The Case of a Point Source

This section presents a continuation of Sect. 3.10. Just like in [13], we consider the SH-wave propagation generated by a point source buried in a hereditary elastic space.

Let x, y, z be Cartesian coordinates in the hereditary space under consideration, r, ψ, z the corresponding cylindrical coordinates $\left(r = \sqrt{x^2 + y^2} \right)$. Let the point source of the rotation center type be located at the origin of coordinates. Then in the Cartesian coordinates, the body force can be represented as

$$\vec{f} = \nabla \times (0, 0, X) \tag{3.11.1}$$

where X is an axisymmetric potential (that is, $X = X(t, r, z)$ in the cylindrical coordinates).

In the Cartesian coordinates, the corresponding field of displacements can be written as [13]

$$\vec{u} = \nabla \times (0, 0, \chi) \tag{3.11.2}$$

where χ also is an axisymmetric potential. From (3.11.2), one easily obtains the cylindrical coordinates of $\vec{u}$

$$u_r = 0, u_\psi = -\frac{\partial \chi}{\partial r}, u_z = 0. \tag{3.11.3}$$

Using (3.11.1, 3.11.2) into Eq. (3.10.1), we arrive at the following wave equation with memory for the potential χ

$$\frac{\partial^2 \chi}{\partial t^2} - \beta_0^2 (1 + h(t)*)\Delta\chi = \frac{X}{\rho} \tag{3.11.4}$$

Here Δ denotes the Laplace operator with respect to x, y, z, and $\beta_0 = (\mu/\rho)^{1/2}$. As in Sect.3.10, we suppose

$$\frac{1}{1 - h(t)*} = 1 + \Lambda(t)*$$

where the creep kernel $\Lambda(t)$ can be represented as

$$\Lambda(t) = \varphi(t) + \frac{1}{4}\varphi(t) * \varphi(t)$$

with $\varphi(t)$ satisfying conditions (a)–(e) from Sect.3.1.

To be specific, just like in [13], we suppose

$$X = A\Theta(t)\delta(x)\delta(y)\delta(z) \equiv A\Theta(t)\frac{\delta(r)}{2\pi r}\delta(z) \tag{3.11.5}$$

where $A = \text{const} \neq 0$

$$\Theta(t) = \begin{cases} 1 & \text{for } t > 0 \\ 0 & \text{for } t < 0 \end{cases}$$

Moreover, we suppose

$$\vec{u} = 0 \; for \; t < 0. \tag{3.11.5'}$$

As in Sect 3.10, our purpose is to construct the near front asymptotics for $u_{y'}$ without applying any integral transformation with respect to z. Just like in the classical approach [13], we apply the multiple transform

$$\int\limits_{-\infty}^{\infty} e^{-i\xi x} \, dx \int\limits_{-\infty}^{\infty} e^{-i\eta y} \, dy \int\limits_{0}^{\infty} e^{-pt} dt$$

to (3.11.4, 3.11.5, 3.11.6). Then we arrive at the ordinary equation

$$\frac{d^2}{dz^2} F_{x,y \to \xi, \eta} \overline{\chi} = -\frac{A}{\rho \beta^2(p)p} \delta(z) + n^2 F_{x,y \to \xi, \eta} \overline{\chi} \tag{3.11.6}$$

where

$$n^2 \equiv \xi^2 + \eta^2 + \frac{p^2}{\beta^2(p)}, \quad \beta(p) = \frac{\beta_0}{1 + \frac{\overline{\varphi}(p)}{2}}.$$

Furthermore, under the restriction $\mathrm{Re}\, n \geq 0$, the solution of (3.11.6) describing finite speed wave propagation obviously has the form

$$F_{x,y \to \xi, \eta} \overline{\chi} = \frac{A}{2\rho\beta^2(p)pn} e^{-n|z|}$$

whence

$$\overline{\chi}(p, x, y, z) = \frac{A}{8\pi^2 \rho \beta^2(p)p} \int\limits_{-\infty}^{+\infty} d\xi \int\limits_{-\infty}^{\infty} \frac{\exp(i\xi x + i\eta y - n|z|)}{n} d\eta. \tag{3.11.7}$$

Now, let us try to asymptotically calculate the right-hand side of Eq. (3.11.7), as $p \to +\infty$. We make use of the following de Hoop's change of variables (see (3.15, 3.16):

$$\xi = (w\cos\psi - q\sin\psi)p, \quad \eta = (w\sin\psi + q\cos\psi)p. \tag{3.11.8}$$

Here ψ is the azimuth in the cylindrical coordinates introduced above in the xyz-space; q and w are new variables introduced instead of ξ and η. Since $x = r\cos\psi$, $y = r\sin\psi$, $d\xi\,d\eta = p^2 dw\,dq$ (3.11.7, 3.11.8) yield

$$\overline{\chi}(p,x,y,z) = \frac{A}{2\pi^2\rho\beta^2(p)} \int_0^\infty dq\,\mathrm{Im} \int_0^{i\infty} \frac{\exp{(ipwr - pN|z|)}}{N}\,dq \tag{3.11.9}$$

where

$$N = \left[\beta^{-2}(p) + q^2 + w^2\right]^{1/2}, \quad \mathrm{Re}\,N > 0.$$

Furthermore, just like in [13], we set $s = -iw$. Then Eq. (3.11.9) assumes the form

$$\overline{\chi}(p,x,y,z) = \frac{A}{2\pi^2\rho\beta^2(p)} \int_0^\infty dq\,\mathrm{Im} \int_0^{i\infty} \frac{\exp{[-p(sr + N|z|)]}}{N}\,ds \tag{3.11.10}$$

where now

$$N = \left[\beta^{-2}(p) + q^2 - s^2\right]^{1/2}, \quad \mathrm{Re}\,N > 0.$$

Now, we recall that in Sect 3.10, it was shown that

$$\mathrm{Im} \int_0^{i\infty} \frac{\exp{[-p(sx + N_1\,|\,z\,|\,]}}{N_1}\,ds = \int_{H/\beta(p)}^\infty \frac{e^{-p\tau}\,d\tau}{\left[\tau^2 - \frac{H^2}{\beta^2(p)}\right]^{1/2}}$$

where

$$N_1 = \left[\beta^{-2}(p) - s^2\right]^{1/2}, \mathrm{Re}\,N_1 > 0; H = \left(x^2 + z^2\right)^{1/2}$$

Replacing here x by $r = (x^2 + y^2)^{1/2}$ and $\beta^{-2}(p)$ by $\beta^{-2}(p) + q^2$, one can easily obtain

$$\text{Im} \int_0^{i\infty} \frac{\exp\left[-p(sr + N|z|)\right]}{N} \, ds = \int_{R\sqrt{\beta^{-2}(p)+q^2}}^{\infty} \frac{e^{-p\tau}d\tau}{\left\{\tau^2 - R^2\left[\beta^{-2}(p) + q^2\right]\right\}^{1/2}}$$

where $R = (x^2 + y^2 + z^2)^{1/2}$. Therefore, Eq. (3.11.10) can be rewritten as

$$\bar{\chi}(p,x,y,z) = \frac{A}{2\pi^2\rho\beta^2(p)} \int_0^\infty dq \int_{R\sqrt{\beta^{-2}(p)+q^2}}^{\infty}$$

$$\times \frac{e^{-p\tau}d\tau}{\left\{\tau^2 - R^2\left[\beta^{-2}(p) + q^2\right]\right\}^{1/2}} \qquad (3.11.11)$$

By interchanging the order of integration in Eq. (3.11.11), one obtains

$$\bar{\chi}(p,x,y,z) = \frac{A}{2\pi^2\rho\beta^2(p)} \int_{R/\beta(p)}^{\infty} d\tau\, e^{-p\tau} \int_0^{\left[\frac{\tau^2}{R^2} - \frac{1}{\beta^2(p)}\right]^{1/2}}$$

$$\times \frac{dq}{\left\{\tau^2 - R^2\left[\beta^{-2}(p) + q^2\right]\right\}^{1/2}} \qquad (3.11.12)$$

Clearly, the external integral in Eq. (3.11.12), which has the type of

$$\int_{G(p,\vec{x})} F\left(p,\tau,\vec{x}\right) e^{-p\tau} \, d\tau, \; \vec{x} = (x,y,z),$$

can be treated as the nonlinear Laplace transform, if one makes the following change of the integration variable

$$\tau \to \frac{G\left(\infty,\vec{x}\right)}{G\left(p,\vec{x}\right)} \tau.$$

Such an approach is rather natural when one is considering the case of a general layered hereditary medium. However, in the particular case of a homogeneous hereditary space, the appropriate integral can be calculated exactly.

In fact, one can easily see that the inner integral in Eq. (3.11.12) equals $\pi/2R$ [13]. Now, by inserting this value of the inner integral into Eq. (3.11.12), one obtains

$$\overline{\chi}(p,x,y,z) = \frac{A}{4\pi R\rho\beta^2(p)} \int\limits_{R/\beta(p)}^{\infty} e^{-p\tau} d\tau = \frac{A}{4\pi R\rho\beta^2(p)p} e^{-pR/\beta(p)}$$

or which is the same,

$$\overline{\chi}(p,x,y,z) = \frac{A\left[1 + \frac{\overline{\varphi}(p)}{2}\right]^2}{4\pi R\mu p} \exp\left\{-p\left[1 + \frac{\overline{\varphi}(p)}{2}\right]\frac{R}{\beta_0}\right\} \tag{3.11.13}$$

Therefore, by virtue of the relation $R = (r^2 + z^2)^{1/2}$, (3.11.3, 3.11.13) yield

$$\overline{u}_\psi = -\frac{\partial}{\partial x}\left(\frac{A\left[1 + \frac{\overline{\varphi}(p)}{2}\right]^2}{4\pi R\mu p} \exp\left\{-p\left[1 + \frac{\overline{\varphi}(p)}{2}\right]\frac{R}{\beta_0}\right\}\right) \tag{3.11.14}$$

$$\sim \frac{Ar}{4\pi R^2 \mu\beta_0} e^{-pR/\beta_0} e^{-p\overline{\varphi}(p)\,R/2\beta_0}, \quad \text{as } p \to +\infty,$$

since $\overline{\varphi}(p) \to 0$, as $p \to +\infty$..

Now, as in Sect.3.10, we consider the special case where

$$\varphi(t) = \frac{kt^{-\alpha}}{\Gamma(1-\alpha)}, \quad \text{for } t > 0. \tag{3.11.15}$$

Here $k > 0$, $0 < \alpha < 1$, Γ is the Gamma function. Then $\overline{\varphi}(p) = kp^{1-\alpha}$, whence (3.11.14) takes on the form

$$\overline{u}_\psi(p,x,y,z) \sim \frac{Ar}{4\pi R^2 \mu\beta_0} e^{-\frac{pR}{\beta_0}} e^{-\frac{p^\alpha kR}{2\beta_0}}, \text{ as } p \to +\infty.$$

Hence

$$u_\psi(t,x,y,z) \sim \frac{Ar}{4\pi R^2 \mu\beta_0} L^{-1}_{p \to t-R/\beta_0} e^{-p^\alpha kR/(2\beta_0)}, \quad \text{as} \quad t - \frac{R}{\beta_0} \to +0. \tag{3.11.16}$$

Therefore, (3.11.16) yields

$$u_\psi(t,x,y,z) \sim \text{const}_1\, rR^{-2+\frac{1}{2(1-\alpha)}} \left(t - \frac{R}{\beta_0}\right)^{-1-\frac{\alpha}{2(1-\alpha)}}$$

$$\times \exp\left[-\text{const}_2 R^{\frac{1}{1-\alpha}} \left(t - \frac{R}{\beta_0}\right)^{-\frac{\alpha}{1-\alpha}}\right] \tag{3.11.17}$$

as $t \to R/\beta_0 + 0$, where

$$\text{const}_1 = \frac{A}{4\pi\mu\beta_0} \frac{\alpha^{\frac{1}{2(1-\alpha)}}}{\sqrt{2\pi(1-\alpha)}} \left(\frac{k}{2\beta_0}\right)^{\frac{1}{2(1-\alpha)}}$$

and

$$\text{const}_2 = (1-\alpha)\,\alpha^{\frac{\alpha}{1-\alpha}}\left(\frac{k}{2\beta_0}\right)^{\frac{1}{1-\alpha}}.$$

Note The contents of Chap. 3 is based on the results of [12, 15–25].

References

1. Lokshin, A. A. and Rok, V.E. *Dokl. Akad. Nauk.* SSSR 239, 1305 (1978).
2. Feller, W.: *An Introduction to Probability Theory and Its Applications*, vol. 2, 2nd edn. Wiley, New York (1971)
3. Widder, D.V.: *The Laplace Transform*. Princeton University Press, Princeton (1946)
4. Seneta, E. (1976). *Regularly Varying Functions*. pp. 1–112. Lecture Notes in Mathematics, Vol. 508 Berlin: Springer.
5. Esseen, G.G.: *Acta Math.* **77**, 1 (1945)
6. Hormander, L.: *Linear Partial Differential Operators*. Springer, Berlin (1963)
7. Follard, H.: *Bull. Amer. Math. Soc.* **10**, 908 (1946)
8. Ibragimov, I.A., Linnik, J.V.: *Independent and Stationarily Linked Random Variables*, pp. 1–524. Nauka, Moscow (1965)
9. Lukacs, E.: *Characteristic Functions*. Griffin, London (1970)
10. Rabotnov, J.N.: *Elements of Hereditary Mechanics of Solids*, pp. 1–383. Nauka, Moscow (1977)
11. Fedoriuk, M.V.: *Asymptotics: Integrals and Series*. Nauka, Moscow (1987)
12. Lokshin, A.A., Sagomonyan, E.A.: *Vestnik Mosc. Univ. ser. Mat. Mekh.* **2**, 87 (1989)
13. Aki, K., Richards, P.: *Quantitative Seismology*, Vol. 1. W. H. Freeman and Company, San Francisco (1980)
14. de Hoop, A.T.: *Appl.Science Research.* **B8, 349** (1960)
15. Lokshin, A. A., Lopatnikov, S. L. and Rok, V. E. *Izv. Akad. Nauk.* SSSR MTT 5, 188 (1990).
16. Lokshin, A.A.: *Prikl, Mat. Mekh.* **1**, 162 (1994)
17. Lokshin, A.A., Suvorova, J.V.: *Mathematical Theory of Wave Propagation in Media with Memory*, pp. 1–151. Moscow University Press, Moscow (1982)
18. Lokshin, A. A. *Dokl. Akad. Nauk.* SSSR 240, 43 (1978).
19. Lokshin, A. A. *Dokl. Akad. Nauk.* SSSR 247, 812 (1979).
20. Lokshin, A. A. *Vestnik Mosc. Univ. Ser, Mat. Mekh.* 2, 93 (1979).
21. Lokshin, A.A.: *Usp. Mat. Nauk.* **1**, 231 (1979)
22. Lokshin, A.A.: *Vestnic Mose.Univ. Ser. Mat. Mech.* **3**, 70 (1979)
23. Lokshin, A.A.: *Vestnik Mosc.Univ. Ser. Mat. Meckh.* **1**, 42 (1981)
24. Lokshin, A. A., Lopatnikov, S. L. and Sagomonyan, E. A. *Izv. Akad. Nauk.* SSSR. Fiz. Zemli 3, 80 (1991).
25. Vinigradova, O. S., Lokshin, A. A. and Rok, V. E. *Izv. Akad. Nauk.* SSSR MTT. 1, 152 (1989).

Appendix: The Near Source Behaviour of Fundamental Solutions for Wave Operators with Memory

In what follows we deal with a specific sort of singularities of fundamental solutions for wave operators with memory. These singularities do not travel along characteristics and remain located at the origin of the space coordinates.

A.1. The One-Dimensional Case

Let $E_1(t, x)$ be the finite speed fundamental solution for the one-dimensional wave operator with memory. That is

$$\frac{\partial^2 E_1}{\partial t^2} - c^2[1 - h(t)*]\frac{\partial^2 E_1}{\partial x^2} = \delta(t)\,\delta(x) \qquad (A.1.1)$$

where $h(t)$ is the relaxation kernel and c the instantaneous elastic velocity. Let $\Lambda(t)$ be the creep kernel corresponding to $h(t)$. Then (A.1.1) is obviously equivalent to

$$[1 + \Lambda(t)*]\frac{\partial^2 E_1}{\partial t^2} - c^2\frac{\partial^2 E_1}{\partial x^2} = [1 + \Lambda(t)*]\,\delta(t)\,\delta(x). \qquad (A.1.1')$$

As in Chap. 3, we suppose

$$\Lambda(t) = \varphi(t) + \frac{1}{4}\varphi(t) * \varphi(t) \qquad (A.1.2)$$

where $\varphi(t)$ satisfies conditions (a)–(e) from Sect. 3.1.

In the manner of Sect. 3.1, one easily obtains the following formula for E_1,

$$E_1(t,x) = F_{\lambda \to t}^{-1} \, \frac{\sqrt{1 + \widetilde{\Lambda}(\lambda)}}{2i\lambda c} \, \exp\left[-i\lambda \sqrt{1 + \widetilde{\Lambda}(\lambda)} \, \frac{|x|}{c}\right], \tag{A.1.3}$$

or, which is the same by virtue of (A.1.2)

$$E_1(t,x) = F_{\lambda \to t}^{-1} \, \frac{1 + \frac{\widetilde{\varphi}(\lambda)}{2}}{2i\lambda c} \, \exp\left\{-i\lambda \left[1 + \frac{\widetilde{\varphi}(\lambda)}{2}\right] \frac{|x|}{c}\right\} \tag{A.1.3$'$}$$

where $\lambda = \mu - ip$, $p > 0$. It can be shown that in the vicinity of $x = 0$ $E_1(t,x)$ is continuous in x for $t > 0$ (see the proof of Theorem 3.3.1). For simplicity, during the whole of the Appendix we suppose $\varphi(t)$ to have a singularity(as $t \to +0$), which is stronger than the logarithmic one. Then, from (A.1.3$'$) and the results of Sect. 3.7, it follows that $E_1(t,x)$ is infinitely differentiable in t, x, for $x \neq 0$. Therefore, (A.1.3) yields

$$\lim_{x \to \pm 0} \frac{\partial}{\partial x} E_1(t,x) = \lim_{x \to \pm 0} F_{\lambda \to t}^{-1} \frac{\partial}{\partial x} \left\{ \frac{\sqrt{1 + \widetilde{\Lambda}(\lambda)}}{2i\lambda c} \, e^{-i\lambda \sqrt{1 + \widetilde{\Lambda}(\lambda)} \frac{|x|}{c}} \right\}$$

$$= \mp F_{\lambda \to t}^{-1} \frac{1 + \widetilde{\Lambda}(\lambda)}{2c^2}$$

whence it follows that

$$\lim_{x \to \pm 0} \frac{\partial}{\partial x} E_1(t,x) = \mp \frac{\Lambda(t)}{2c^2}, \quad \text{for} \, t > 0. \tag{A.1.4}$$

Thus, we have found a specific singularity of $E_1(t,x)$ which is located at $x = 0$.

A.2. The Three-Dimensional Case

Let $E_3(t,x,y,z)$ be a finite speed fundamental solution for a three-dimensional wave operator with memory. That is

$$\frac{\partial^2 E_3}{\partial t^2} - c^2[1 - h(t)*]\Delta E_3 = \delta(t)\,\delta(x)\,\delta(y)\,\delta(z), \tag{A.2.1}$$

or

$$[1 + \Lambda(t)*]\frac{\partial^2 E_3}{\partial t^2} - c^2\Delta E_3 = [1 + \Lambda(t)*]\,\delta(t)\,\delta(x)\,\delta(y)\,\delta(z), \qquad (A.2.1')$$

Here Δ is the Laplace operator with respect to x, y, $0z$; $h(t)$ and $\Lambda(t)$ are the relaxation and creep kernel, respectively; c is the instantaneous elastic velocity for the appropriate type of waves. By applying the Fourier–Laplace transform to (A.2.1') and transition to spherical coordinates, one can easily obtain

$$E_3(t,x,y,z) = F_{\lambda\to t}^{-1}\frac{1 + \widetilde{\Lambda}(\lambda)}{4\pi Rc^2}\,e^{-i\lambda\sqrt{1+\widetilde{\Lambda}(\lambda)}(R/c)} \qquad (A.2.2)$$

where $R = (x^2 + y^2 + z^2)^{1/2}$. Since E_3 proves to be a spherically symmetric function, in what follows, we denote it as $E_3(t,R)$.

From (A.1.4), (A.2.2), it follows that

$$E_3(t,R) = -\frac{1}{2\pi R}\frac{\partial}{\partial R}E_1(t,R) \qquad (A.2.3)$$

Hence (A.1.4) yields

$$E_3(t,R) \sim \frac{\Lambda(t)}{4\pi Rc^2}, \quad \text{as } R \to +0, t > 0, \qquad (A.2.4)$$

provided $\Lambda(t) \neq 0$, for $t > 0$.

A.3. The Two-Dimensional Case

Now, let $E_2(t,x,y)$ be a finite speed fundamental solution for a two-dimensional wave operator with memory. That is

$$\frac{\partial^2 E_2}{\partial t^2} - c^2[1 - h(t)*]\,\Delta E_2 = \delta(t)\,\delta(x)\,\delta(y), \qquad (A.3.1)$$

or

$$[1 + \Lambda(t)*]\frac{\partial^2 E_2}{\partial t^2} - c^2\Delta E_2 = [1 + \Lambda(t)*]\,\delta(t)\,\delta(x)\,\delta(y). \qquad (A.3.1')$$

Here Δ is the Laplace operator with respect to x, y.

We are going to construct the fundamental solution $E_2(t,x,y)$ by the Hadamard's method of descent. That is

$$E_2(t, x, y) = \int\limits_{-\infty}^{+\infty} E_3\left[t, \sqrt{x^2 + y^2 + (z - \xi)^2}\right] d\xi$$

$$= \int\limits_{-\infty}^{\infty} E_3\left(t, \sqrt{x^2 + y^2 + z^2}\right) dz = \int\limits_{-\infty}^{\infty} E_3(t, R) \, dz$$

$$= \int\limits_{R \leq ct} E_3(t, R) \, dz$$

$$= 2 \int\limits_{0}^{\sqrt{(ct)^2 - r^2}} E_3(t, R) \, dz \qquad (A.3.2)$$

where $R = \sqrt{x^2 + y^2 + z^2}, r = \sqrt{x^2 + y^2}$. Let us introduce the following change of variables in (A.3.2)

$$z = \sqrt{R^2 - r^2}, \quad r = \text{const.}$$

Then, on account of the relation

$$dz = \frac{R \, dR}{\sqrt{R^2 - r^2}},$$

Eqs. (A.3.2) and (A.2.3) yield

$$E_2(t, x, y) = 2 \int\limits_{r}^{ct} \frac{R E_3(t, R)}{\sqrt{R^2 - r^2}} \, dR = -\frac{1}{\pi} \int\limits_{r}^{ct} \frac{1}{\sqrt{R^2 - r^2}} \frac{\partial}{\partial R} E_1(t, R) \, dR. \qquad (A.3.3)$$

In accordance with Eq. (A.3.3), we denote E_2 as $E_2(t, r)$. Then, from Eq. (A.3.3) and (A.1.4), it easily follows that

$$E_2(t, r) \frac{\tilde{\Lambda}(t)}{2\pi c^2} \ln r, \qquad r \to +0, t > 0, \qquad (A.3.4)$$

provided $\Lambda(t) \neq 0$, for $t > 0$.

Note From Eqs. (A.1.4), (A.2.4), and (A.3.4), it is clear that wave operators with memory possess certain properties of the Laplace operators.

Printed in the United States
by Baker & Taylor Publisher Services

Printed in the United States
by Baker & Taylor Publisher Services